Transactions on Computer Systems and Networks

Series Editor

Amlan Chakrabarti, Director and Professor, A.K.Choudhury School of Information Technology, Kolkata, West Bengal, India

Transactions on Computer Systems and Networks is a unique series that aims to capture advances in evolution of computer hardware and software systems and progress in computer networks. Computing Systems in present world span from miniature IoT nodes and embedded computing systems to large-scale cloud infrastructures, which necessitates developing systems architecture, storage infrastructure and process management to work at various scales. Present day networking technologies provide pervasive global coverage on a scale and enable multitude of transformative technologies. The new landscape of computing comprises of self-aware autonomous systems, which are built upon a software-hardware collaborative framework. These systems are designed to execute critical and non-critical tasks involving a variety of processing resources like multi-core CPUs, reconfigurable hardware, GPUs and TPUs which are managed through virtualisation, real-time process management and fault-tolerance. While AI, Machine Learning and Deep Learning tasks are predominantly increasing in the application space the computing system research aim towards efficient means of data processing, memory management, real-time task scheduling, scalable, secured and energy aware computing. The paradigm of computer networks also extends it support to this evolving application scenario through various advanced protocols, architectures and services. This series aims to present leading works on advances in theory, design, behaviour and applications in computing systems and networks. The Series accepts research monographs, introductory and advanced textbooks, professional books, reference works, and select conference proceedings.

More information about this series at https://link.springer.com/bookseries/16657

Narendra Kumar · Celia Shahnaz · Krishna Kumar ·
Mazin Abed Mohammed · Ram Shringar Raw
Editors

Advance Concepts of Image Processing and Pattern Recognition

Effective Solution for Global Challenges

 Springer

Editors
Narendra Kumar
School of Computing
DIT University
Dehradun, India

Krishna Kumar
Research and Development Unit
UJVN Ltd.
Dehradun, India

Ram Shringar Raw
Department of Computer Science
and Engineering
Netaji Subhash University of Technology
(East Campus)
New Delhi, India

Celia Shahnaz
Department of Electrical and Electronic
Engineering
Bangladesh University of Engineering
and Technology
Dhaka, Bangladesh

Mazin Abed Mohammed
College of Computer Science
and Information Technology
University of Anbar
Ramadi, Iraq

ISSN 2730-7484   ISSN 2730-7492 (electronic)
Transactions on Computer Systems and Networks
ISBN 978-981-16-9323-6   ISBN 978-981-16-9324-3 (eBook)
https://doi.org/10.1007/978-981-16-9324-3

This Springer imprint is published by the registered company Springer Nature Singapore Pte Ltd.
The registered company address is: 152 Beach Road, #21-01/04 Gateway East, Singapore 189721,
Singapore

Preface

Image processing is one of the challenging fields of engineering. Various applications of image processing are being used in the field of medical imaging, cyberforensic, satellite imagery, smart agriculture, etc. Our aim is to explain the important concepts and principles of image processing so that readers can easily implement the algorithms and techniques and lead themselves to discover new problems and applications. This book will discuss numerous fundamental and advanced image processing algorithms and pattern recognition techniques to illustrate the framework.

The main strength of this book is it will present essential background theory, shape methods, texture about new methods, and techniques for image processing and pattern recognition. We have made a good balance between a mathematical background and practical implementation. This book also contains the comparison table and images that are used to show the results of enhanced techniques. This book consists of novel concepts and hybrid methods for providing effective solutions for society. It will also include a detailed and explanatory of algorithms in various programming languages like MATLAB, Python, etc.

This book is new and special for those people who are working in the field of image processing, pattern recognition, and security for digital images. The security features of image processing like image watermarking and image encryption, etc., are available in this book. This book is helpful for academicians and researchers to provide complete knowledge with the help of theoretical concepts, mathematical background and illustrated the procedure of image processing and pattern recognition.

Readers

This book is helpful for researchers, academicians, and developers working in the areas of image processing, pattern recognition, medical imaging, and telemedicine.

The main features of the book are:

- It has covered all the latest developments and future aspects of image processing.

- This book is very useful for the new researchers working in the field to quickly know-how the best performing methods.
- The book is concisely written, lucid, comprehensive, application-based, graphical, schematics and covers wider aspects of image processing.

Chapter Organization

Chapter 1: This chapter presents an overview of contrast enhancement techniques and has proposed one hybrid technique based on an evolutionary algorithm. The proposed technique has been applied to many low-contrast images, and the performance has been compared with the cuckoo search technique, artificial bee colony technique, and other similar techniques.

Chapter 2: In this chapter, the issues related to classification and identification of rice grains are of critical significance at the industrial level for the manufacturing and packaging foodstuffs has been discussed. This study focused on the development of the visual inception system for Sri Lankan rice grain quality based on features of the rice grains.

Chapter 3: Presented an image enhancement and noise reduction technique.

Chapter 4: Discussed the pattern recognition problem and its various stages in detail. In addition to that, the application of deep learning and machine learning in pattern recognition has also been explained.

Chapter 5: Discussed the brain tumor classification using a hybrid artificial neural network with chicken swarm optimization algorithm which consists of various pre-processing and thresholding to extract the lung parenchyma.

Chapter 6: Explained the design of a ConvNet model to extract features from the histological image of breast cancer based upon obtained relevant features by the same network contrary to the current model.

Chapter 7: In this chapter, a partial differential equation filter is introduced to restore noisy images based on anisotropic diffusion-based method in the L-2 framework.

Chapter 8: This chapter aims to discuss the state-of-the-art systems for detecting diabetic eye diseases with traditional and deep learning techniques. A statistical comparison is also made using various performance metrics.

Chapter 9: Proposed a hybrid method to reduce speckle noise from ultrasound images while maintaining a tradeoff between speckle reduction and edge preservation. Extensive work is carried out by taking four different types of images for research: synthetic images, simulated images, noise-free ultrasound images, and original ultrasound images.

Chapter 10: A diversity of approaches for de-noising mammogram pictures have been described, each with its own assumptions, advantages, and limitations. The efficacy of the filters has also been compared using criteria such as MSE and PSNR.

Chapter 11: This chapter, presented image enhancement technique by utilizing the morphological technique. The arrangement of elements determined the increase of binary pixels in an image. When erosion is applied to the greyscale image, it enhances an image's intensity by enchanting the region extreme after fleeting the constructing part over the object. Outcomes of the technique carried out to several sides of an object indicate that abstract data of the reference object can be effortlessly obtained.

Chapter 12: In this chapter, a despairing depiction-based estimation is used to figure the affirmation cost and offers information to situate evaluation.

Dehradun, India   Narendra Kumar
Dhaka, Bangladesh   Celia Shahnaz
Dehradun, India   Krishna Kumar
Ramadi, Iraq   Mazin Abed Mohammed
New Delhi, India   Ram Shringar Raw

Contents

About the Editors

Narendra Kumar is a Faculty at the Department of School of Computing, DIT University, Dehradun, India. He is M.Tech. (Computer Science) from BIT Mesra, Ranchi, Jharkhand, and Ph.D. (Computer Science) from Deen Dayal Upadhyaya Gorakhpur University, Gorakhpur. He has over 12 years of teaching experience. He has published numerous research papers in international journals and conferences. His research areas include data science, IoT, and image processing.

Celia Shahnaz, SMIEEE, Fellow IEB, received a Ph.D. degree from Concordia University, Canada, and is currently a Professor at the Department of EEE, BUET, Bangladesh, since 2015. She has published over 150 international journal/conference papers. She is a recipient of the Canadian Commonwealth Scholarship/Fellowship and Bangladesh Academy of Science Gold Medal for her contribution to Science and Technology. Her research interests include the areas of deep learning, pattern recognition and machine learning for audio, video, biomedical and power signals, signal processing for speech analysis and speech enhancement, multimodal emotion recognition, digital watermarking, audio-visual recognition for biometric security, multimedia communication, control system, robotics, and humanitarian technology. She has around 20 years of experience (over 18 years as an IEEE volunteer) in leading impactful technical, professional, educational, industrial, women empowerment, and humanitarian technology projects at national/international levels.

Krishna Kumar is a Research and Development Engineer at UJVN Ltd. (A Government of Uttarakhand Enterprises). Before joining UJVNL, he has worked as Assistant Professor at BTKIT, Dwarahat. He has received his B.E. degree from Govind Ballabh Pant Engineering College, Pauri Garhwal, and M.Tech. from Motilal Nehru NIT, Allahabad. He is pursuing his Ph.D. from the Indian Institute of Technology, Roorkee. He has over 11 years of experience with numerous research papers published in international journals and conferences. He has also edited several books published by International publishers. His present research areas include IoT, machine learning, image processing, and renewable energy.

Mazin Abed Mohammed obtained his B.Sc. from the University of Anbar, Iraq, M.Sc. from the College of Graduate Studies, Universiti Tenaga Nasional (UNITEN), Malaysia. and a Ph.D. degree from Universiti Teknikal Malaysia Melaka, Malaysia. He has published over 60 research papers in journals, book chapters, conferences, and tutorials. Dr. Mohammed is the Editor in Chief of *Journal of Fusion: Practice and Applications*, Associate Editor of *International Journal of Smart Sensor Technologies and Applications*IGI Global: International Publisher of Information Science and Technology Research. He is a reviewer of over 40 reputed Journals. His specialization and research interests include artificial intelligence, biomedical computing and processing, medical image and data processing, machine learning, deep learning, optimization methods, and software medical IoT.

Ram Shringar Raw received his Ph.D. from Jawaharlal Nehru University, New Delhi, India. He has obtained his M.Tech. (IT) and B.E. (CSE) in 2005 and 2000 respectively. He has worked as an Associate Professor in the Department of Computer Science of Indira Gandhi National Tribal University (A Central University, Madhya Pradesh) from April 2016 to March 2018. He is currently working as an Associate Professor in the Department of Computer Science and Engineering of Netaji Subhas University of Technology, East Campus, Delhi, India. He has over 18 years of teaching, administrative and research experience. Dr. Raw has published over 100 research papers. He has supervised several B.Tech., M.Tech., and Ph.D. students for their dissertation and thesis work. His current research interest includes Mobile Ad hoc Networks, Vehicular Ad hoc Networks, Flying Ad-hoc Networks, and Cloud Computing.

Chapter 1
Hybrid Evolutionary Technique for Contrast Enhancement of Color Images

Narendra Kumar, Krishna Kumar, Anil Kumar Dahiya, and Rachna Shah

Abstract Image contrast enhancement is an important task in image processing, which intends to restore the true image from degraded images. The images may be degraded due to noise or blur, and image contrast enhancement performs tasks like deblurring, denoising, etc. Digital images are very essential and important part of our everyday life. The enhancement of contrast is an effective step in image processing as well as in computer vision applications. The histogram equalization technique is used to enhance the contrast of the image, but the histogram equalization technique is not popular due to the artificial effect, immense brightness change, and over enhancement. In this paper, we have proposed a hybrid technique for image contrast enhancement based on an evolutionary algorithm. The proposed technique has been applied to many low-contrast images, and the performance has been compared with the Cuckoo search technique, artificial bee colony technique, and other similar techniques. The nonlinear models are well-optimized with the aid of bio-inspired optimization algorithms to operate adaptively with the noise and blurring models. The proposed approach has shown adequate image enhancement with average results of 26.11 as PSNR score, 620.43 MSE, 24.85 RMSE, 0.5943 UQI, 205.44 MAE, 0.4445 NAE, 4.2992 entropy, and 3.7053 mutual information. Hence, the developed algorithm will be a good compromise to operate between contrast enhancement and denoising techniques.

Keywords Evolutionary technique · Contrast enhancement · Image processing · Denoising

N. Kumar · A. K. Dahiya
School of Computing, DIT University, Dehradun, Uttarakhand, India

K. Kumar (✉)
Department of Hydro and Renewable Energy, IIT Roorkee, Roorkee, India

R. Shah
National Informatics Centre, Dehradun, Uttarakhand, India

1.1   Introduction

The uses of a digital image are growing continuously, and video and television transmission become more digital. The low-light image enhancement algorithm aims at generating visually appealing images and collecting useful details for applications in computer vision. One challenging task is to enhance the quality of low-light images (Srinivas and Bhandari 2020). To correct the color of the picture according to the characteristics of each display, global histogram equalization is applied. Local dual-interval histogram equalization based on the average of peak and mean values is used to boost image contrast according to each channel's characteristics (Bai et al. 2020). When image size is low, the picture may lose more essential details. The image is therefore transformed from spatial to wavelet domain to achieve multi-resolution (Mahmood et al. 2019; Shuka et al. 2014). Image brightness for grayscale images is related to gray. Gray means the average brightness of the image indicated by the value. If it does not shine, the image is too dark, and there is nothing we can see clearly from the image (Shukla et al. 2014; Kumar et al. 2017). This also makes people feel awkward when the picture is too vivid (Li et al. 2019). A fuzzy-based approach that uses a membership function depends on the image mean intensity value (Kumar et al. 2017). The details of the image level and the low contrast are inconsistent with the visual perception of humans. This type of problem may be caused by variations in the environment or by limitations on the camera. A multi-scale top-hat transform-based method improves image entropy and image detail level improvement (Román et al. 2019).

Swarm intelligence is a family of algorithms derived from the behavior pattern of "natural swarms." Swarm stands for "a large group of people or things" as per the online Oxford dictionary. In nature, insects, ants, bees, and birds live in swarms. The study of behavioral patterns of these natural swarms has shown that the swarm behavior outperforms the sum of an individual behavior (Kumar et al. 2017). Alternately collective gains of efforts of a swarm exceed the sum of the individual efforts. In other words, individuals need to make less effort in the swarm to achieve the same goal. In a "swarm," all individuals contribute toward solutions of a common goal without any centrally coordinated efforts. Swarm behavior is a good example of cooperative learning. A good example would be long flights of bird flocks. Individual birds in such a flock tend to position in a formation in which flying movements of leading birds ease efforts required from the others. To prevent the tiring of any bird, the responsibility of leading the flock is shifted periodically. This helps the entire flock to reach the destination in a shorter time or cover larger distances. The particle swarm optimization method is derived from the behavior of a flock of birds. Both position and velocity are deriving factors of the optimization method. Many bionic-based algorithms have been developed for image enhancement like bat algorithm, cuckoo algorithm, artificial ant colony algorithm, etc. Multi-objective bat and neuron networks are used to enhance the contrast of images. The enhanced cuckoo search algorithm is used to enhance the contrast of grayscale images, and by using a new range of search

space, it is used to optimize the four local/global parameters enhancement transformations (Kamoona and Patra 2019). The histogram equalization approach consists
of three stages like histogram analysis, image brightness levels clustering, and individual cluster contrast enhancement (Shakeri et al. 2017). Artificial bee colony (ABC)
has been developed for solving the multi-objective problem (Al-Ammar et al. 2020).
The artificial bee colony method has given a positive effect on various applications
like segmentation of medical images, clustering, image enhancement, and image
classification (Öztürk and Akhtar 2020). The adaptive cuckoo search algorithm has
been used to optimize bilateral filter for image denoising and for preserving the
edges of images (Asokan and Anitha 2020). Cuckoo search algorithm and discrete
wavelet transform (DWT) have been used to improve contrast, and CS algorithm has
been applied to each sub-band of DWT optimization and then achieved the matrix
with the singular value of the low threshold sub-band image (Bhandari et al. 2014).
The contrast enhancement technique with the help of cuckoo search and multi-scale
retinex algorithm has been also used for microscopic image quality improvement
(Biswas et al. 2015). Artificial bee colony-based multilevel threshold has been used
for iris detection which is helpful in the unification of the iris region (Bouaziz et al.
2015). Artificial bee colony algorithm has been also used for the optimization of
fitness function and generate new pixel intensities for enhanced images with help of
transformation function (Draa and Bouaziz 2014). In the cuckoo search algorithm,
usage of biased random walk and the population reduction are the most frequently
used modifications (Jamil and Zepernick 2013).

The contribution of this study can be summarized into the following points:

- Propose a hybrid technique for image contrast enhancement based on artificial
  bee colony and cuckoo search algorithms.
- A quantitative as well as qualitative assessment based on different measurements
  and different evolutionary techniques are presented.

1.2   Evolutionary Techniques Background

1.2.1   Artificial Bee Colony (ABC) Technique

The ABC method is a metaheuristic swarm-based algorithm, which is used to solve
the problems of mathematical optimization. The ABC is inspired by the action of
the natural honeybee. The population of artificial bee colonies is divided into three
subparts, such as working bees, onlooker bees, and scout bees. Such artificial bees
choose the food sources by moving in search space, which is the possible solution
for the goal optimization problem. It is the responsibility of working bees and the
onlooker bees who are responsible for sharing the knowledge to research different
food sources. The onlooker bees choose a higher score and source of higher quality
food. On the other hand, there is less risk that food sources with lower marks would
be picked. Because of their poorer quality, the low-quality food source may also be

refused. In this case, the scout bees must perform a random search for new food sources. Therefore, each search iteration process involves three steps:

(i)   The employee bees are sent to food source search and measure the food quality.
(ii)   The onlooker selects the food sources after information sharing with employees bees.
(iii)   For searching for new possible food sources, the scout bees are sent.

1.2.2 Cuckoo Search Algorithm (CSA)

This population-based metaheuristic algorithm was aspired by the reproduction behavior of cuckoo birds. The following algorithm describes the concepts of the cuckoo search as shown in Figs. 1.1 and 1.2.

```
Begin
    Objective function f(X), X = (X₁, X₂, ...,Xd)ᵀ ;
    Initial population of n host nests Xᵢ (i = 1, 2, ..., n);
    while (t <MaxGeneration) or (stop criterion)
        Get a Cuckoo randomly by Lévy flights;
        Evaluate its quality/fitness Fᵢ;
        Choose a nest among n (say, j) randomly;
    if (Fᵢ < Fⱼ)
            Replace j by the new solution;
        end
        A fraction (Pₐ) of worse nests are abandoned and new ones are built;
        Keep the best solutions (or nests with quality solutions);
        Rank the solutions and find the current best;
    end while
End
```

1.3 Proposed Hybrid Image Contrast Enhancement Technique

The proposed technique is based on the combination of artificial bee colony and cuckoo search algorithm.

Step 1:   Extract out the red, green, and blue pixel matrixes from the color image.
Step 2:   Obtain the histogram and un-normalized discrete probability mass function of pixel intensities for each matrix.
Step 3:   Set the parameters, lambda, the quantity for positioning the amount of contrast on a scale of 0–20, and gamma, the amount of detail in the image to be retained, on a scale from $1-10^9$. Usually, lambda is around 10, and gamma is 50,000.
Step 4:   For each pixel matrix, obtain the optimized histogram from the cuckoo search optimization algorithm. Cuckoo search will take arguments as

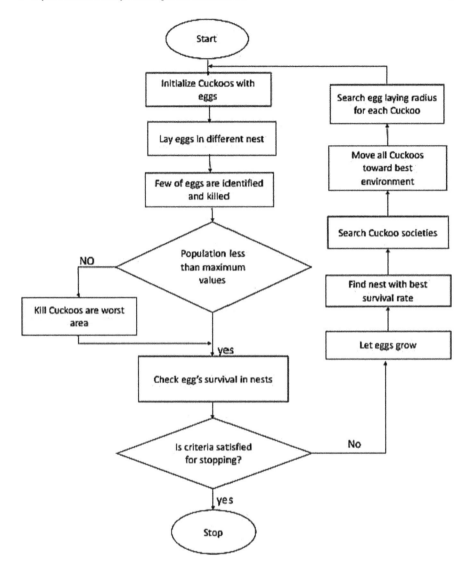

Fig. 1.1 Flow chart of cuckoo search algorithm (CSA)

the number of nests, input histogram, hi[n], lambda, gamma, and the difference matrix.

The histogram equalization is defined as for individual sub-histogram is

$$X_{i-1} + (X_i - X_{i-1}) \cdot CDF_i$$

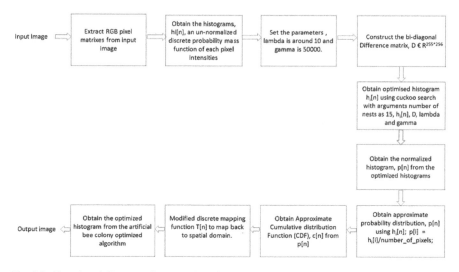

Fig. 1.2 Functional diagram of proposed work

where X_i and X_{i-1} are the lower and higher end of the ith sub-histogram that is dynamic. The CDF_i is the cumulative Distribution function (CDF) of ith sub-histogram.

Step 5:   Obtain the normalized histograms, $p[n]$, from the optimized histograms. It gives an approximate probability distribution function of the pixel intensities.

Step 6:   Then, the approximate CDF, $c[n]$, is obtained from $p[n]$.

Step 7:   After the CDF is obtained, a modified discrete mapping function $T[n]$ is used to map back to the spatial domain (pixels).

$$T[n] = (\lambda + 1) \times \{(2^{\beta} - 1) \times (\text{sum}(p[1:n]) + 0.5)\}$$

where β is the number of bits used to represent the pixel values and $n \in [0, 2^{\beta}-1]$ and $p[n]$ is the probability density function.

Step 8:   Obtain the optimized histogram from the artificial bee colony optimization algorithm. This algorithm will take arguments as the number of nests, input histogram, lambda, gamma, and the difference matrix of image that is obtained from mapping.

Step 9:   Go to step 5, step 6, and step 7.

Step 10:   Finally, the image is obtained from this mapping. After this, the required performance metrics are calculated like PSNR, entropy, mean, etc.

Fig. 1.3 Original image

1.4 Results and Discussion

The performance of proposed techniques has been analyzed with performance metrics like PSNR, MSE, MAE, etc. Figure 1.3 shows the original images of size 256×256.

The color threshold is used to identify the object of consistent color values. The red, green, and blue histograms are displayed by the interface. Figure 1.4 shows the RGB threshold. The single-pixel color value is the combination of the individual value of RGB (red, green, and blue) channels. Figure 1.5 shows the RGB threshold of contrast images. Figures 1.6, 1.7, and 1.8 show the RGB threshold of existing contrast enhancement techniques.

1.5 Image Quality Measurement

The performance of the proposed technique has been measured based on parameters like PSNR, MSE, RMSE, UQI, and MAE. The peak signal-to-noise ratio (PSNR) is the proportion of an image's maximum possible power to the power of corrupting

Fig. 1.4 Original image-RGB threshold

noise that affects its representation efficiency. To calculate an image's PSNR, it is important to compare it to an ideal clean image with the highest possible power.

$$\text{PSNR} = 10 \log_{10}\left(\frac{(L-1)^2}{\text{MSE}}\right) \tag{1.1}$$

Here, L is the number of maximum possible intensity levels.
MSE is the mean squared error, and it is defined as:

$$\text{MSE} = \frac{1}{mn}\sum_{i=0}^{m-1}\sum_{j=0}^{n-1}(O(i,j) - D(i,j))^2 \tag{1.2}$$

The root mean square error (RMSE) is given as the squared root of MSE. The root mean square error (RMSE) measures the amount of change per pixel due to the processing.

$$\text{RMSE} = \sqrt{\text{MSE}} \tag{1.3}$$

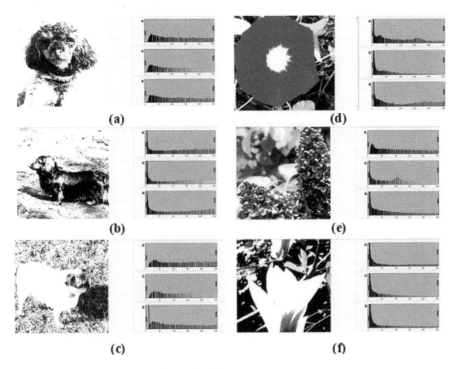

Fig. 1.5 Proposed techniques-RGB threshold

Image quality index (IQI) is a universal or normal measure of pixel variations between two images (image1 and image2), with a value ranging from −1 to 1. If both images are similar, the IQI is 0. If the IQI is less than 1 but similar to 1 (say, 0.8704), the enhanced image's output is said to be higher. The best value 1 is achieved if and only if $y_i = x_i$ for all $i = 1, 2, \ldots, N$. The lowest value of -1 occurs when $y_i = 2\overline{x} - x_i$ for all $i = 1, 2, \ldots N$. The definition of Q is a product of three components.

$$Q = \frac{\sigma_{xy}}{\sigma_x \sigma_y} \frac{2\overline{x}\,\overline{y}}{(\overline{x})^2(\overline{y})^2} \frac{2\sigma_x \sigma_y}{\sigma_x^2 \sigma_y^2} \tag{1.4}$$

The mean absolute error (MAE) is a statistic that measures the difference in errors between paired observations describing the same phenomenon. Comparisons of expected versus observed, subsequent time versus initial time, and one measurement technique versus another measurement technique are examples of Y versus X. The MAE is measured as follows:

$$\text{MAE} = \frac{\sum_{i=1}^{n} |y_i - x_i|}{n} \tag{1.5}$$

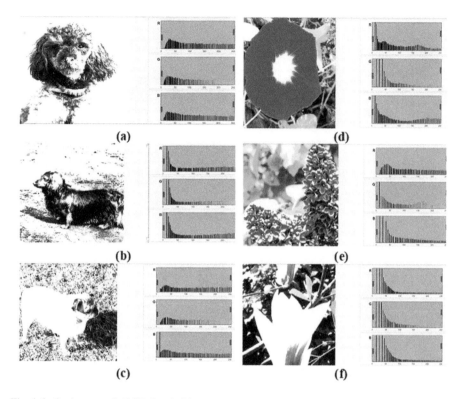

Fig. 1.6 Cuckoo search-RGB threshold

The proposed technique gives a higher PSNR value than other existing techniques for all images as shown in Table 1.1 and Fig. 1.9.

The proposed technique gives a lower MSE value than other existing techniques for all images as shown in Table 1.2 and Fig. 1.10.

The proposed technique gives a lower RMSE value than other existing techniques for all images as shown in Table 1.3 and Figs. 1.10 and 1.11.

The proposed technique gives a higher UQI value than other existing techniques for all images as shown in Table 1.4 and Fig. 1.12.

The proposed technique gives a lower MAE value than other existing techniques for all images as shown in Table 1.5 and Fig. 1.13.

1.6 Image Error Measurement

Normalized absolute error (NAE) calculation can be used to measure the performance of the proposed technique. The proposed technique gives a lower NAE value than other techniques for all the images as shown in Table 1.6 and Fig. 1.14.

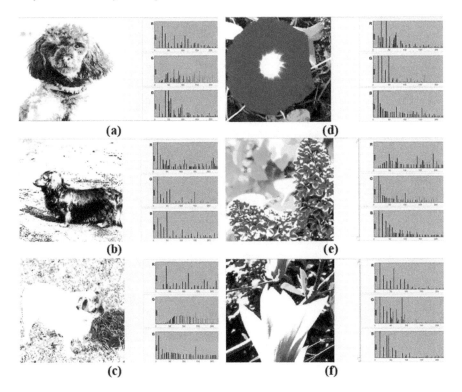

Fig. 1.7 PSO-RGB threshold

$$NAE = \frac{\sum_{i=1}^{M} \sum_{j=1}^{N} \text{abs}(I_{\text{ref}} - I_{\text{est}})}{\sum_{i=1}^{M} \sum_{j=1}^{N} \text{abs}(I_{\text{ref}})} \tag{1.6}$$

The entropy of average information of an image is a measure of the degree of randomness in the image.

$$\text{Entropy, } H = -\sum_{k} p_k \log_2(p_k) \tag{1.7}$$

where k is the number of gray levels and P_k is the probability associated with gray level k.

The proposed technique gives a higher information value than other techniques for all the images as shown in Table 1.7 and Fig. 1.15.

Mutual information (MI) is an effective similarity measure for comparing images. Mutual information is closely related to joint entropy. Specifically, given image A and image B, the joint entropy H (A, B) can be calculated as:

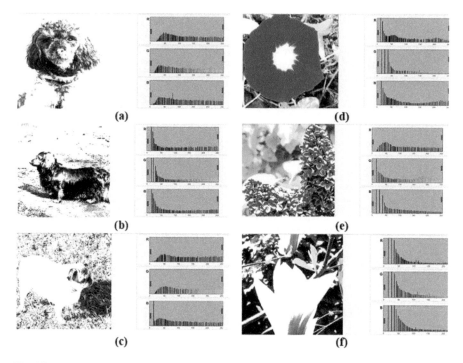

Fig. 1.8 WDO-RGB threshold

Table 1.1 Peak signal-to-noise ratio (PSNR)

Image	PSNR				
	PSO	WDO	Cuckoo search	Artificial bee colony	Proposed technique
(a)	24.07547088	24.06912	24.17532	24.23205165	24.2812
(b)	24.9989002	25.1293	25.36285	24.93286168	25.54473
(c)	24.81136388	25.09354	25.33971	24.80250572	25.48703
(d)	27.35599898	26.6727	26.84437	25.92001772	27.36991
(e)	25.7572242	25.95679	26.16774	25.29591275	27.11537
(f)	26.82958506	26.40473	26.4517	25.75014837	26.90502

$$H(A, B) = - \sum_{a,b} p_{AB(a,b)} \log p_{AB(a,b)} \qquad (1.8)$$

The proposed technique gives a higher mutual information value than other techniques for all the images as shown in Table 1.8 and Fig. 1.16.

PSNR(Peak Signal to Noise Ratio)

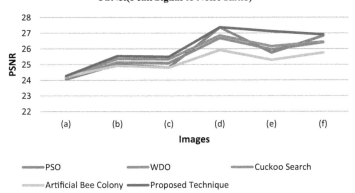

Fig. 1.9 Noise removal performance of the proposed technique and existing techniques for all the images

Table 1.2 Mean square error (MSE) of the proposed technique and existing techniques for all the images

Image	MSE				
	PSO	WDO	Cuckoo search	Artificial bee colony	Proposed technique
(a)	727.7845	762.6091	744.612	734.5574	729.7413177
(b)	762.3859	760.8569	749.0281	732.2466	650.6403046
(c)	757.4241	760.2396	747.7776	732.6264	707.5356293
(d)	750.1585	753.0751	746.5687	684.13	564.4811249
(e)	759.8466	760.1148	738.4082	715.048	581.3668671
(f)	728.6909	728.6909	716.858	667.7453	488.8226013

MSE

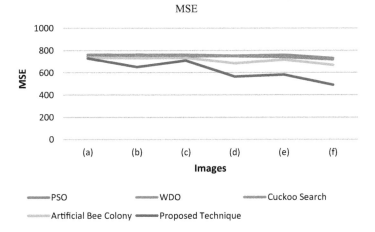

Fig. 1.10 MSE comparisons of the proposed technique and existing techniques for all the images

Table 1.3 Root mean square error (RMSE) of the proposed technique and existing techniques for all the images

Image	RMSE				
	PSO	WDO	Cuckoo search	Artificial bee colony	Proposed technique
(a)	26.9775	27.6154	27.2876	27.1027	27.0137
(b)	27.6113	27.5836	27.3684	27.0601	25.5077
(c)	27.5213	27.5724	27.3455	27.0671	26.5995
(d)	27.389	27.4422	27.3234	26.1559	23.7588
(e)	27.5653	27.5702	27.1737	26.7404	24.1116
(f)	26.9943	26.9943	26.7742	25.8408	22.1093

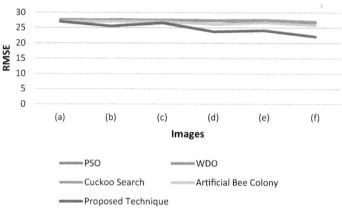

Fig. 1.11 RMSE of the proposed technique and existing techniques for all the images

Table 1.4 Universal quality index of the proposed technique and existing techniques for all the images

Image	Universal quality index (UQI)				
	PSO	WDO	Cuckoo search	Artificial bee colony	Proposed technique
(a)	0.720771654	0.624987031	0.672066502	0.59427389	0.699212958
(b)	0.512403677	0.526081486	0.561443491	0.4942997	0.617536266
(c)	0.510148044	0.547501164	0.59040927	0.507681957	0.627616348
(d)	0.481136138	0.419889014	0.437274249	0.383622355	0.60233442
(e)	0.32057473	0.33437418	0.353704177	0.281476233	0.532036439
(f)	0.542128962	0.484576601	0.490280672	0.434375405	0.487632609

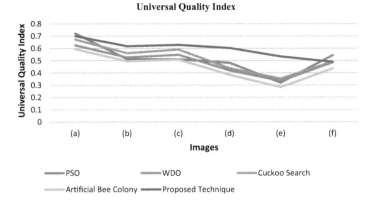

Fig. 1.12 Universal quality index of the proposed technique and existing techniques for all the images

Table 1.5 Mean absolute error of the proposed technique and existing techniques for all the images

Image	Mean absolute error (MAE)				
	PSO	WDO	Cuckoo search	Artificial bee colony	Proposed technique
(a)	219.6136475	265.9817352	241.4506226	302.447113	232.1205292
(b)	307.5178833	291.8388367	265.7134857	322.0767517	250.258667
(c)	341.6659698	302.3894501	274.7513275	346.6839905	249.246521
(d)	209.4853363	240.6582336	227.7261963	287.4515076	185.0928955
(e)	313.0726471	296.0232086	275.955719	387.4757538	175.2225342
(f)	159.1805725	197.912796	192.8401337	262.7599335	140.7379456

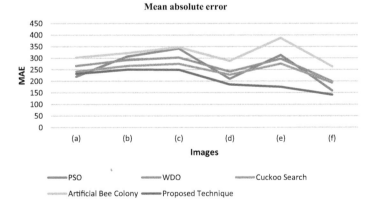

Fig. 1.13 MAE of the proposed technique and existing techniques for all the images

Table 1.6 Normalized absolute error

Image	Normalized absolute error (NAE)				
	PSO	WDO	Cuckoo search	Artificial bee colony	Proposed technique
(a)	0.413059856	0.431003277	0.407317662	0.462222629	0.395617707
(b)	0.489472678	0.485034479	0.477479663	0.500493527	0.480117669
(c)	0.52104866	0.516143609	0.516639845	0.536102289	0.515941486
(d)	0.342851896	0.348536385	0.343887318	0.366840598	0.343190569
(e)	0.4254461	0.413404818	0.404409608	0.457538541	0.380417011
(f)	0.511022948	0.520360112	0.520463898	0.531253162	0.551826498

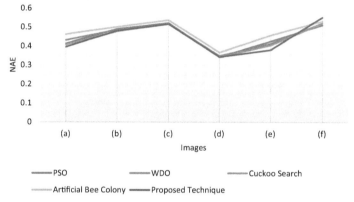

Fig. 1.14 MAE of the proposed technique and existing techniques for all the images

Table 1.7 Entropy of original image

Image	Entropy of the original image	Entropy of the contrast image				
		PSO	WDO	Cuckoo search	Artificial bee colony	Proposed technique
(a)	7.777937608	3.249358202	3.278512876	3.612348685	2.010892527	3.617565608
(b)	7.622879288	2.718845645	3.265756145	3.729235991	2.208842985	3.687924568
(c)	7.848369831	2.733136634	3.785123623	4.137826866	2.216776677	4.353006949
(d)	7.179074832	3.782161628	4.206503668	4.352088512	3.035442377	4.364238074
(e)	7.247414768	4.49055067	4.890292125	5.268634101	2.826785869	5.767185917
(f)	7.038816655	3.733328325	3.907747826	4.308921634	2.943696882	4.005738373

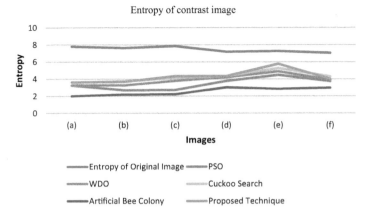

Fig. 1.15 Entropy of the proposed technique and existing techniques for all the images

Table 1.8 Mutual information of the proposed technique and existing techniques for all the images

Image	Mutual information				
	PSO	WDO	Cuckoo search	Artificial bee colony	Proposed technique
(a)	2.563981555	2.816150296	3.097683687	1.739964961	3.122903083
(b)	2.264602522	2.783221392	3.180212926	1.947266804	3.367794616
(c)	2.242530756	3.200524515	3.509180973	1.860138107	3.692824557
(d)	2.967016511	3.501088375	3.60463529	2.511609808	3.739659706
(e)	3.759326107	4.086969451	4.342419231	2.372631916	4.722915292
(f)	3.192058871	3.386069845	3.467371933	2.418766465	3.686211457

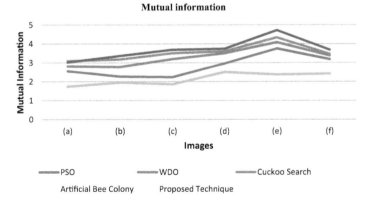

Fig. 1.16 Mutual information of the proposed technique and existing techniques for all the images

1.7 Conclusion

In this paper, we have reviewed some of the potential papers to show the image quality enhancement techniques. Artificial bee colony and cuckoo search algorithm gives good results, and the proposed hybrid technique for image contrast enhancement gives higher PSNR value than any individual technique. The proposed technique is a hybrid technique based on the artificial bee colony and cuckoo search algorithm. The quantitative as well as qualitative assessment with metrics like entropy, PSNR, MSE, RMSE, IQI, MAE, etc., are used to evaluate the performance of the proposed technique. Mainly four popular techniques namely PSO, WDO, cuckoo search, and artificial bee colony are used as the baseline for comparison with the proposed approach. The proposed approach has shown adequate image enhancement with average results of 26.11 as PSNR value, 620.43 MSE, 24.85 RMSE, 0.5943 UQI, 205.44 MAE, 0.4445 NAE, 4.2992 entropy, and 3.7053 mutual information. The PSNR value of the proposed technique is higher, and MSE is less as compared to the other techniques. The performance of the proposed hybrid technique is better than other existing techniques.

References

Al-Ammar EA et al (Jun. 2020) ABC algorithm based optimal sizing and placement of DGs in distribution networks considering multiple objectives. Ain Shams Eng J. https://doi.org/10.1016/j.asej.2020.05.002

Asokan A, Anitha J (2020) Adaptive cuckoo search based optimal bilateral filtering for denoising of satellite images. ISA Trans 100:308–321

Bai L, Zhang W, Pan X, Zhao C (2020) Underwater image enhancement based on global and local equalization of histogram and dual-image multi-scale fusion. IEEE Access 8:128973–128990. https://doi.org/10.1109/access.2020.3009161

Bhandari A, Soni V, Kumar A, Singh G (2014) Cuckoo search algorithm based satellite image contrast and brightness enhancement using DWT–SVD. ISA Trans 53(4):1286–1296

Biswas B, Roy P, Choudhuri R, Sen B (2015) Microscopic image contrast and brightness enhancement using multi-scale retinex and cuckoo search algorithm. Proc Comput Sci 70:348–354. https://doi.org/10.1016/j.procs.2015.10.031

Bouaziz A, Draa A, Chikhi S (2015) Artificial bees for multilevel thresholding of iris image. Swarm Evolution Comput 21:32–40. https://doi.org/10.1016/j.swevo.2014.12.002

Draa A, Bouaziz A (2014) An artificial bee colony algorithm for image contrast enhancement. Swarm Evolution Comput 16:69–84. https://doi.org/10.1016/j.swevo.2014.01.003

Jamil M, Zepernick H (2013) Multimodal function optimisation with cuckoo search algorithm. Int J Bio-Inspired Comput 5(2):73. https://doi.org/10.1504/ijbic.2013.053509

Kamoona AM, Patra JC (2019) A novel enhanced cuckoo search algorithm for contrast enhancement of gray scale images. Appl Soft Comput J 85. https://doi.org/10.1016/j.asoc.2019.105749

Kumar N, Shukla HS, Tripathi RP (2017) Image restoration in noisy free images using fuzzy based median filtering and adaptive particle swarm optimization—richardson-lucy algorithm. Int J Intell Eng Syst 10:50–9. https://doi.org/10.22266/ijies2017.0831.06

Kumar N, Kumar A, Kumar DK (2020) Image restoration using a fuzzy-based median filter and modified firefly optimization algorithm. Int J Adv Sci Technol 29:1471–14777

Kumar N, Kumar A, Kumar DK (2020) Modified median filter for image denoising. Int J Adv Sci Technol 29:1495–502

Li C, Liu J, Liu A, Wu Q, Bi L (2019) Global and adaptive contrast enhancement for low illumination gray images. IEEE Access 7:163395–163411. https://doi.org/10.1109/ACCESS.2019.2952545

Mahmood A, Khan SA, Hussain S, Almaghayreh EM (2019) An adaptive image contrast enhancement technique for low-contrast images. IEEE Access 7:161584–161593. https://doi.org/10.1109/ACCESS.2019.2951468

Öztürk RA, Akhtar N (2020) Variants of artificial bee colony algorithm and its applications in medical image processing. Appl Soft Comput 97:106799

Román JCM, Noguera JLV, Legal-Ayala H, Pinto-Roa DP, Gomez-Guerrero S, Torres MG (2019) Entropy and contrast enhancement of infrared thermal images using the multiscale top-hat transform. Entropy 21:1–19. https://doi.org/10.3390/e21030244

Shakeri M, Dezfoulian MH, Khotanlou H, Barati AH, Masoumi Y (2017) Image contrast enhancement using fuzzy clustering with adaptive cluster parameter and sub-histogram equalization. Digit Signal Process A Rev J 62:224–237. https://doi.org/10.1016/j.dsp.2016.10.013

Shuka H, Kumar N, Tripathi PR (2014) Median filter based wavelet transform for multilevel noise. Int J Comput Appl 107:11–4. https://doi.org/10.5120/18818-0225

Shukla H, Kumar N, Tripathi RP (2014) Gaussian noise filtering techniques using new median filter. Int J Comput Appl 95:12–5. https://doi.org/10.5120/16645-6617

Srinivas K, Bhandari AK (2020) Low light image enhancement with adaptive sigmoid transfer function. IET Image Process 14:668–678. https://doi.org/10.1049/iet-ipr.2019.0781

Chapter 2
Computer Vision for Agro-Foods: Investigating a Method for Grading Rice Grain Quality in Sri Lanka

H. M. K. K. M. B. Herath, G. M. K. B. Karunasena, and R. D. D. Prematilake

Abstract Rice is the most important food crop in developing or developed countries and the staple diet of more than half of the world's population. It is high in nutrients, minerals, and a good source of complex carbohydrates. When compared to wheat and maize, human consumption accounted for 78% of global rice production in 2009. Recent technological advancements have brought attention to the importance of modern vision technology in improving rice quality. Classification and identification of rice grains are of critical significance at the industrial level for the manufacturing and packaging of foodstuffs. That is why machine vision technologies are creating a new path in the classification and analysis of rice grains in industry. This study focused on the development of the visual inception system for estimating Sri Lankan rice grain quality based on features of the rice grains. The proposed machine learning methodology achieved 91.8% accuracy, while the overall system showed 90.31% accuracy. Based on the experimental findings, the proposed system showed promising results by classifying Sri Lankan rice grains.

Keywords Image classification · Machine learning · Machine vision · Feature extraction · Rice Analysis

2.1 Introduction

Sri Lankan history of rice production or paddy cultivation traces its origins back to the proud history of 161 B.C. and in A.D. 1017. Rice (*Oryza sativa L.*) is the single most significant crop in Sri Lanka, accounting for 34% of the total cultivated area. On average, 560,000 ha are cultivated during Maha and 310,000 ha are cultivated during Yala, allowing an average annual region of around 870,000 ha of rice. Approximately, 1.8 million farm families are involved in paddy production throughout the country. Sri Lanka reportedly generates 2.7 million tons of rice each year and meets about

H. M. K. K. M. B. Herath (✉) · G. M. K. B. Karunasena
Faculty of Engineering Technology, The Open University of Sri Lanka, Nugegoda, Sri Lanka

R. D. D. Prematilake
Faculty of Science and Technology, University of Plymouth, Plymouth, UK

N. Kumar et al. (eds.), *Advance Concepts of Image Processing and Pattern Recognition*, Transactions on Computer Systems and Networks, https://doi.org/10.1007/978-981-16-9324-3_2

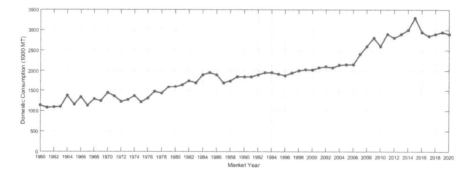

Fig. 2.1 Domestic rice consumption in Sri Lanka since 1960

95% of domestic requirements. Rice provides an average of 45% of total calories and 40% of the total protein needs of humans (Weerakoon et al. 2011). The demand for rice is estimated to increase by 1.1% per year and to fulfill rice supply; it is anticipated to increase by 2.9% per year. Increased crop size and national average yields are the strategies necessary to meet such development goals. The United Nations Food and Agriculture Organization's (UNFAO) census recorded that the overall production of rice in the world has risen from 570 million tons in 2002 to 720 million tons in 2012 (Statistics, F. A. O 2010). Figure 2.1 indicates the domestic consumption of Sri Lankan rice over time and demand has increased by 300,000 MT over the last 10 years.

Deep learning has been increasingly established in recent years (Son et al. 2019), such as search engines, recommendation systems, image recognition, and voice recognition systems (LeCun et al. 1989). Greater results have been obtained by convolutional neural networks (CNNs) for image recognition (Hijazi et al. 2015). Support vector machine (SVM) (Ibrahim et al. 2019; Nagoda and Ranathunga 2018) can be used for purposes of classification (Hamzah and Mohamed 2020; Koklu et al. 2021) and regression. These factors have benefited the agricultural sector's development in recent years. Advances in agricultural technologies are a significant contribution to a more productive agricultural system that encourages quality development and less resource usage.

Image processing is an evolving technology for the rapid growth of the agro-foods industry. Through the application of visual imagery and computer videos, artificial vision has been commonly employed in several specific areas in recent years. Many approaches for identifying rice varieties have been suggested based on biological or chemical strategies, such as genetic markers (Cirillo et al. 2009). Spatial features and spectral features are the primary extraction methods used by optical imaging systems. Different spatial features defining the visual appearance of the rice seeds include shape, morphological, and textural features (Haralick et al. 1973). An examination of the image based on the shape, morphology, and color characteristics of the grains is important in this field. The morphological characters of the grains are genetic in nature and play a significant role in the recognition of the variety. The image processing techniques aim to classify different rice grains, and the consistency of the rice grains is dependent on a variety of parameters such as grain color, shape, length, and width (Herath and Mel 2017; Karunasena and Priyankara 2020).

Rice quality must be maintained due to the growing worldwide demand for rice and rice-based food products. This study is intended to provide an effective machine learning approach for the classification of rice grains using machine vision techniques. The following is a summary of the remainder of this chapter: The related works are described in Sect. 2.2, and the methodology is described in Sect. 2.3. Section 2.4 discusses the findings and observations. Finally, in Sect. 2.5, we bring the study to a conclusion.

2.2 Literature Review

Using Google Scholar, and ResearchGate, we looked at a number of research studies on rice grain classification and its applications. This section discusses the different machine learning techniques and deep learning techniques used to classify rice grains in previous studies.

Pratibha et al. (2017) have used image recognition and neural networks to investigate and categorize rice grains. They also built a rating method to simplify labor-intensive work and to maintain continuity in the standard of the goods. Their program is helpful in categorizing granule grades utilizing neural network pattern recognition. This method has focused on the extraction features of rice granules. Features derived from the image of the rice granules have area, perimeter, major axis, and minor axis.

Silva and Upul (2013) have developed a separate artificial neural network model for the individual features set and the unified features set. High classification accuracy has provided by textural features as well as morphological and color features. The average classification precision of 92.00% has been achieved from the integrated function model.

The rice varieties have been identified using the spatial–spectral information of the two types of datasets gathered by Chatnuntawech et al. (2018) using a deep CNN algorithm. The proposed approach achieved a mean classification accuracy of

Table 2.1 Rice varieties used for the experiments

Rice type	Code	Samples size
White basmathi rice	OSL1	100
White samba rice	OSL2	100
White sudu kekulu rice	OSL3	100
Red kekulu rice	OSL4	100
Red basmathi rice	OSL5	100
Red raw rice	OSL6	100

91.09% for the paddy rice dataset. In their study, all rice seeds have examined using a hyperspectral imaging device of the same orientation.

Kaur and Singh (2013) have suggested a machine learning algorithm for grading rice kernels using multi-class SVM. In their study, the maximum variance approach has used to remove rice kernels from the background, and then, the chalk extracted from rice. Multi-class SVM has used to recognize the rice kernels by analyzing the shape, chalkiness, and percentage of the damaged kernels. The SVM classification has shown more than 86.00% reliability. Based on the findings, it has been established that the method is adequate to identify and label the various varieties of rice grains based on their internal and external quality.

Chaugule and Mali (2014) have analyzed the texture and shape features using artificial neural network architectures for the classification of cereal grains. An evaluation of the precision of the classification of texture and form characteristics and the neural network has been performed to identify four paddy rice grains. The accuracy has reached 82.61%, 88.00%, and 87.27%, respectively, with texture, shape, and texture–shape features.

Pazoki et al. (2014) have carried out an experiment at the Islamic Azad University in 2011 to identify five major varieties of rice grain produced in various ecosystems in Iran. The classification has rendered on the basis of 24 color characteristics, 11 morphological characteristics, and 4 shape factors derived from the color images of the growing grain of rice. The rice grains have been categorized by multi-layer perceptron (MLP) and neuro-fuzzy neural networks according to variety. The topological framework of the MLP model comprised 39 input layer neurons, 5 output layer neurons, and 2 hidden layers; with 60 laws, the neuro-fuzzy classifier implemented the same framework in input and output layers. The overall quality rates for the classification of rice grain varieties have 99.46% and 99.73%, respectively, for MLP and neuro-fuzzy classifiers.

Anami et al. (2015) have proposed a system for distinguishing paddy varieties from specimens of bulk paddy grain images based on surface texture characteristics derived from surface matrices. The color texture features and their variations have obtained from the color planes H, S, and I. The feature collection has been limited dependent on feature input to the precision of the recognition. The paddy grain images have been classified using a multilayer feed-forward artificial neural network, and 92.33% accuracy is achieved for 15 bulk paddy grain samples.

A research team (Aki et al. 2015) from the University of Trakya have carried out a study on the classification of rice grains using visual recognition and machine learning techniques. There are three types of grain have used for sorting including broken grains. Rice grain images have been obtained from a Web camera. For every grain object, six attributes have been extracted. The attributes of recognized varieties of rice grains have been used for the training of several machine learning algorithms. Many of the closest outputs of the generalization algorithm neighbors have been chosen for real-time analysis. The specificity of the test outcome is measured as 90.50% with the selected classifier.

Anusha and Shabari (2016) have proposed a mechanism for rice grain classification obtained by image analysis with the design of neural networks. Experiments have been performed with 20 sample images of grain, and each contained 50–100 grains of rice. Reference images have been used to create the data sets for the input function. The feedforward pattern recognition neural network is used to train algorithms once the training function dataset and target dataset were developed. A neural network pattern recognition system has been used for the lucrative grading of rice granules. The developed neural network has been adapted with 91.30% accuracy to recognize and identify rice grains.

The next section of this study describes the materials and methodology of the proposed system.

2.3 Materials and Methods

The image acquisition process was performed with a digital camera from Logitech C270. The camera was mounted on a standard lighting panel on a fixed frame. In this study, a black surface was used as a backdrop and images were taken on the black backdrop by scattering the grains. Six varieties of rice grains were used to conduct the experiment. The traditional Sri Lankan rice samples used for this study are depicted in Table 2.1.

Figure 2.2 illustrates the rice grading parameters based on the kernel length. As shown in the illustration, rice kernels were graded (long, medium, short, and broken) according to the length of the kernel.

Figure 2.3 illustrates the architecture of the proposed system. As shown in the illustration, the proposed system was developed with preprocessing, image segmentation, morphological feature extraction, CNN platform, feature classification, and parameter extraction process. Finally, rice grading was performed based on the data extracted from the feature classification and parameter extraction process.

The grading process was performed according to the parameters explained in Table 2.2. Once the color and morphological features were extracted (feature classification process, and parameters extraction process), the grading was performed.

The image was converted to grayscale after being captured in three-dimensional RGB color space (Karunachandra and Herath 2021) (red, green, and blue). Grayscale images (Yoshioka et al. 2007) are made up of pixels with intensities ranging from 0 to 255. The grayscale image was then converted to a binary image, which only comprises black (binary value 0) and white pixels (binary value 1). The threshold

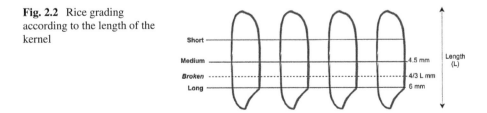

Fig. 2.2 Rice grading according to the length of the kernel

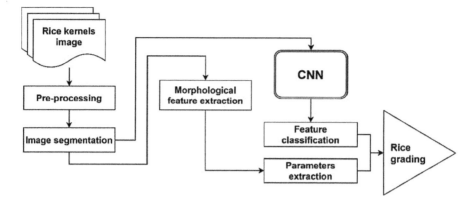

Fig. 2.3 Architecture of the proposed rice grading framework

Table 2.2 Sri Lankan rice grading parameters

Type of rice length	Length
Long red	> 6.0 mm
Long white	> 6.0 mm
Medium red	4.5–6.0 mm
Medium white	4.5–6.0 mm
Short red	< 4.5 mm
Short white	< 4.5 mm
Broken rice	< 3/4th of grain length in mm

function was applied to the grayscale image to generate the binary image. We were able to exclude rice grain regions from the primary background for feature extraction due to the extreme thresholding. Additionally, grain regions were recovered from the background by selecting pixels with a binary value of 1. Figure 2.4 illustrates the proposed preprocessing model in this study.

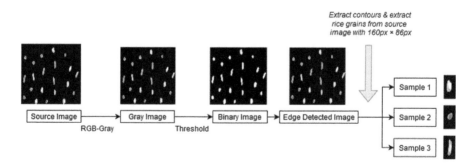

Fig. 2.4 Architecture of the proposed preprocessing procedure

In most cases, an image will include certain areas that are unimportant. As a result, segmentation occurs, in which the image is separated into several sections and only the Region of Interest (ROI) is analyzed. This makes it easier to find the necessary objects and limits in an image. Many approaches, such as region-based segmentation, edge-based segmentation, clustering-based segmentation (Manoharan 2020), etc., can be used to achieve segmentation. In the binary converted image, the edge detection algorithm was utilized to locate the borders of rice grains. The object contour is the most important need in shape analysis, object identification, and recognition processes since the contour reveal the object border; we extracted features of rice grains utilizing these contours.

The proposed CNN model for color separation is illustrated in Fig. 2.5. One input layer, two convolutional layers, two pooling layers, three fully connected layers, a Softmax layer, and a classification layer were comprised of the proposed model.

Table 2.3 depicts the specifications of each layer in the proposed CNN model. The input layer describes 168 pixels of image height, 86 pixels of image width, and 3 for RGB color values, as shown in the table. Each convolution layer consisted of 2 × 2 filters. Each convolution layer was connected with a pooling layer with 2 × 2 filters. Finally, three fully connected layers were employed, each containing 500, 100, and 2 neurons.

After each convolution layer, it is customary to add a nonlinear layer. The objective of this layer is to provide nonlinearity to a system where the convolution layers have

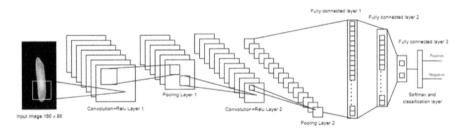

Fig. 2.5 Architecture of the proposed CNN model

Table 2.3 Specifications of the proposed CNN model

Layer	Parameters
Input	168 × 86 × 3
Convolution layer 1	2 × 2 filters
Pooling layer 1	2 × 2 filters
Convolution layer 2	2 × 2 filters
Pooling layer 2	2 × 2 filters
Fully connected Layer 1	500 neurons
Fully connected Layer 2	100 neurons
Fully connected Layer 3	2 neurons

primarily calculated linear operations. ReLU layers are considerably more effective since they allow the network to learn more quicker without losing accuracy. It also aids in overcoming the vanishing gradient problem, which occurs when the gradient falls exponentially down the layers, causing the bottom levels of the network to train extremely slowly. All of the values in the input volume were subjected to the ReLU layer's function $f(x) = \max(0, x)$. This layer was simply restored all negative activations to 0. Without influencing the convolution layer's receptive fields, this layer was enhanced the model's and overall network's nonlinear properties.

We were used a pooling layer after applying some ReLU layers (Chen et al. 2020). A down-sampling layer is another name for it. There are numerous layer choices in this area, with Maxpooling being the most common. This consisted of a filter (2×2) and a stride of the same length. It then applied it to the input volume and produced the highest number in each of the filter's subregions.

The Softmax function converts a vector of K real values into a vector of K real values that sum to 1. In here, the Softmax was converted the input values into values between 0 and 1, allowing them to be recognized as probabilities. The input values might be positive, negative, zero, or higher than one. The Softmax converted a small or negative input into a small probability and a high input into a large probability, but it always stayed between 0 and 1. Finally for multi-class classification problems with mutually exclusive classes, a classification layer was used to compute the cross-entropy loss.

For the training process, the following configurations were used in the MATLAB code editor.

```
'InitialLearnRate', 0.1
'LearnRateDropFactor', 0.01
'LearnRateDropPeriod', 8, ...
'L2Regularization', 0.1, ...
'MaxEpochs', 20
'MiniBatchSize', 10
```

The CNN testing methodology employed in this study is depicted in Fig. 2.6. Preprocessing was done before testing the framework. Images were cropped to 168 $\times$ 86 pixels before being sent to the CNN testing model. The CNN models 1 to 6 represent the features of six different rice grain varieties.

The method then used the specified functions from the binary converted picture to discover the morphological characteristics (Macalalad et al. 2019) of each rice grain. Images of rice grains were used to extract morphological characteristics such as area, major axis length (L), and minor axis length (W). The distance between the ends of the longest line was called maximum length (major axis length). The pixel distance between every combination of pixels in the grain contour (Kuo et al. 2016) was used to calculate the main axis length. The next section of this study briefly describes the experimental findings on each test trial.

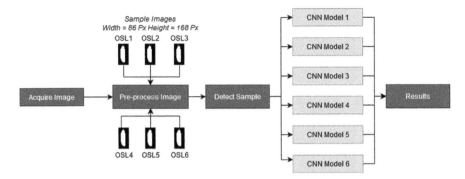

Fig. 2.6 Architecture of the CNN testing model

2.4 Results

In this section, we discuss the experiment results and analysis of each experiment. We performed ten experiment trials on six types of rice varieties that are available in Sri Lanka. Each testing sample is composed of 20% of grains from the original rice grain sample of each kind. Testing was performed using the Logitech C270 Web camera device. Table 2.4 depicts the results of ten experiment trials. Each experiment was recorded the probability value of each rice variety.

Figure 2.7 illustrates the confusion matrix results for rice classification based on color features.

Equations (2.1) and (2.2) describe the accuracy and precision of experimental results that derived from the confusion matrix.

$$\text{Accuracy (CNN}_A) = \frac{\sum(TP + TN)}{\sum(TP + FP + FN + TN)} \tag{2.1}$$

$$\text{Precision (CNN}_P) = \frac{\sum TP}{\sum(TP + FP)} \tag{2.2}$$

Table 2.4 CNN test results for ten samples

Grain type	Experiment									
	1	2	3	4	5	6	7	8	9	10
OSL1	1.0	0.9	0.9	0.9	1.0	1.0	0.9	0.8	0.9	0.8
OSL2	0.9	0.9	1.0	1.0	0.9	0.8	0.9	0.8	1.0	1.0
OSL3	0.9	1.0	0.9	0.8	0.9	0.9	0.8	0.9	0.9	0.9
OSL4	0.8	1.0	0.8	1.0	0.9	0.8	0.9	0.7	0.9	0.8
OSL5	0.9	0.9	0.8	0.9	0.9	0.9	1.0	0.9	0.9	0.9
OSL6	1.0	0.8	0.9	0.9	0.9	0.9	0.9	0.9	0.8	0.9

Fig. 2.7 Confusion matrix
of the CNN prediction

		Predicted value					
		OSL1	OSL2	OSL3	OSL4	OSL5	OSL6
True value	OSL1	9	1	0	0	0	0
	OSL2	0	10	0	0	0	0
	OSL3	0	1	9	0	0	0
	OSL4	0	0	0	8	1	1
	OSL5	0	0	0	0	9	1
	OSL6	0	0	0	0	1	9

where

TN   The number of correct predictions of a negative case,
TP   The number of correct predictions of a positive case,
FP   The number of incorrect predictions of a positive case,
FN   The number of incorrect predictions of a negative case.

According to the experimental results, the overall accuracy (CNN_A) of the CNN prediction was found to be 91.80% at 0.902 precision (CNN_P). According to Fig. 2.7, OSL2 was detected with the highest accuracy. The system successfully classified rice grains based on the color features. Also, experimental results suggested that there was no detection of opposite-colored rice grains.

The resulting window of the rice grain analysis framework is shown in Fig. 2.8. Green-tagged rice kernels were recognized as medium white, as displayed in the window. Following CNN's classification of color features, the rice analysis framework was classified them based on morphological characteristics such as width and height. The majority of rice kernels in this picture had an average length of 5.67 mm, which was classified as medium white. Table 2.5 shows the results of three trials of the experiment.

Figure 2.9 illustrates the confusion matrix of the three experiments. As shown in the illustration, the diagonal represents the highest values compared to other cells.

The maximum possible error for false detection was 0.15. We observed that the Mean Absolute Error (MAE) for each experiment was lower than 0.2. Therefore, the analysis suggested that the proposed system worked with the highest efficiency.

Table 2.6 depicts the true/positive (N_{TP}), accuracy, and precision values of each experiment. According to the test results, the average accuracy of the proposed system was achieved 90.31% at 0.9096 precision.

2.5   Conclusion

Rice is the most essential food crop in developing and developed countries, and more than half of the world's population consumes it as a staple food. Computer vision technologies are now commonly used in various sectors of agricultural production and industrial food production. Rice is the world's most often consumed food; thus,

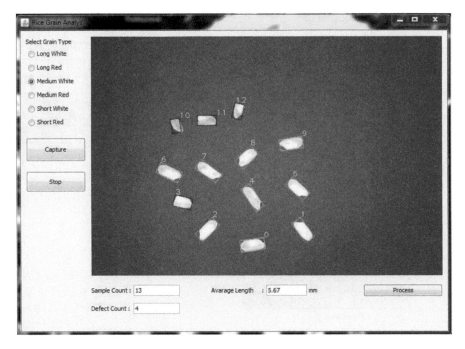

Fig. 2.8 Result window of the rice grain analysis platform

Table 2.5 Experiment results on three trials

Rice type	Trial 1	Trial 2	Trial 3
Long white	0.95	0.95	0.90
Long red	0.90	0.95	0.95
Medium white	0.90	0.90	0.85
Medium red	0.80	0.85	0.90
Short white	0.90	0.95	0.85
Short red	0.90	0.90	0.95

it is essential to use modern technology to improve rice quality. This study focused on the rice varieties available in Sri Lanka. We proposed a rice grading system based on color and rice kernel size. A total of 600 rice kernels were used in the experiment. Twenty percent (20%) of each original sample was used in the testing. The Logitech C270 Web camera was used for the image acquisition process. According to the test results, the overall accuracy (CNN_A) of the CNN prediction was found to be 91.80% at 0.902 precision (CNN_P). The accuracy of the grading system was 90.31% at 0.9096 precision. In conclusion, the proposed system produced positive results in terms of classifying Sri Lankan rice. The next step for the authors is to propose an improved framework for estimating rice quality in Sri Lanka. During the COVID-19 (Herath et al. 2021) period in Sri Lanka, this method can be used to estimate the quality of rice

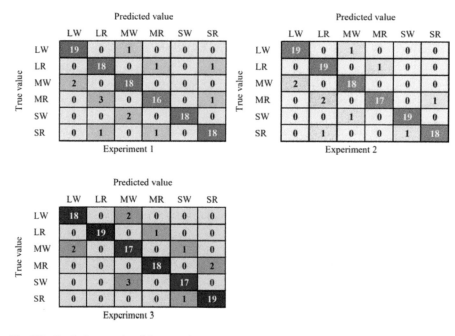

Fig. 2.9 Confusion matrix of the experiments

Table 2.6 True/positive (N_{TP}), accuracy, and precision of each experiment

Experiment no.	N_{TP}	Accuracy (%)	Precision
1	107	89.17	0.916
2	110	91.67	0.913
3	108	90.10	0.900

for buyers. We may need to estimate rice quality using the remote method because the virus is transmitted through direct contact with the infected person's respiratory droplets. The authors will utilize IoT technologies (Herath 2021) to develop the system.

Conflict of Interest   The authors declare that they have no conflicts of interest.

References

Aki O, Güllü A, Uçar E (2015) Classification of rice grains using image processing and machine learning techniques

Anami B, Malvade N, Hanamaratti N (2015) Behavior of HSI color co-occurrence features in variety recognition from bulk paddy grain image samples. Int J Signal Process Image Process Pattern Recogn 8:19–30

Anchan A, Shedthi S (2016) Classification and Identification of rice grains using neural network. Int J Innov Res Comput Commun Eng 4(4):5160–5167

Chatnuntawech I, Tantisantisom K, Khanchaitit P, Boonkoom T, Bilgic B, Chuangsuwanich E (2018) Rice classification using spatio-spectral deep convolutional neural network. arXiv preprint arXiv: 1805.11491

Chaugule A, Mali SN (2014) Evaluation of texture and shape features for classification of four paddy varieties. J Eng 2014:1–8

Chen J, Zhang D, Nanehkaran YA, Li D (2020) Detection of rice plant diseases based on deep transfer learning. J Sci Food Agric 100(7):3246–3256

Cirillo A, Del Gaudio S, Di Bernardo G, Galderisi U, Cascino A, Cipollaro M (2009) Molecular characterization of Italian rice cultivars. Eur Food Res Technol 228(6):875–881

Hamzah AS, Mohamed A (2020) Classification of white rice grain quality using ANN: a review. IAES Int J Artif Intell 9(4):600

Haralick RM, Shanmugam K, Dinstein IH (1973) Textural features for image classification. IEEE Trans Syst Man Cybern 6:610–621

Herath HMKKMB (2021) Internet of Things (IoT) enable designs for identify and control the COVID-19 pandemic. In: Artificial intelligence for COVID-19. Springer, Cham, pp 423–436

Herath K, de Mel WR (2017) Rice grains classification using image precessing technics. Int J Sci Eng Res 10–14

Herath HMKKMB, Karunasena GMKB, Herath HMWT (2021) Development of an IoT based systems to mitigate the impact of COVID-19 pandemic in smart cities. In: Machine intelligence and data analytics for sustainable future smart cities. Springer, Cham, pp 287–309.

Hijazi S, Kumar R, Rowen C (2015) Using convolutional neural networks for image recognition. Cadence Design Systems Inc., San Jose, pp 1–12

Ibrahim S, Zulkifli NA, Sabri N, Shari AA, Noordin MRM (2019) Rice grain classification using multi-class support vector machine (SVM). IAES Int J Artif Intell 8(3):215

Karunachandra RTHSK, Herath HMKKMB (2021) Binocular vision-based intelligent 3-D Perception for robotics application. Int J Sci Res Publ 10:689–696

Karunasena GMKB, Priyankara H (2020) Tea bud leaf identification by using machine learning and image processing techniques. Int J Sci Eng Res

Kaur H, Singh B (2013) Classification and grading rice using multi-class SVM. Int J Sci Res Publ 3(4):1–5

Koklu M, Cinar I, Taspinar YS (2021) Classification of rice varieties with deep learning methods. Comput Electron Agric 187:106285

Kuo TY, Chung CL, Chen SY, Lin HA, Kuo YF (2016) Identifying rice grains using image analysis and sparse-representation-based classification. Comput Electron Agric 127:716–725

LeCun Y, Boser B, Denker JS, Henderson D, Howard RE, Hubbard W, Jackel LD (1989) Backpropagation applied to handwritten zip code recognition. Neural Comput 1(4):541–551

Macalalad CL, Arboleda ER, Andilab AA, Dellosa RM (2019) Morphological based grain comparison of three rice grain variety. Int J Sci Technol Res 8(08):1446–1450

Manoharan S (2020) Performance analysis of clustering-based image segmentation techniques. J Innov Image Process (JIIP) 2(01):14–24

Nagoda N, Ranathunga L (2018) Rice sample segmentation and classification using image processing and support vector machine. In: 2018 IEEE 13th international conference on industrial and information systems (ICIIS). IEEE, pp 179–184

Pazoki AR, Farokhi F, Pazoki Z (2014) Classification of rice grain varieties using two artificial neural networks (mlp and neuro-fuzzy). J Anim Plant Sci 24:336–343

Pratibha N, Hemlata M, Krunali M, Khot ST (2017) Analysis and identification of rice granules using image processing and neural network. Int J Electron Commun Eng 10(1):25–33

Silva C, Upul S (2013) Classification of rice grains using neural networks. In: Proceedings of technical sessions, vol 29, Sri Lanka

Son NH, Thai-Nghe N (2019) Deep learning for rice quality classification. In: 2019 international conference on advanced computing and applications (ACOMP). IEEE, pp 92–96

Statistics, F. A. O. (2010). Food and agriculture organization of the United Nations. Retrieved 3(13):2012

Weerakoon WMW, Mutunayake MMP, Bandara C, Rao AN, Bhandari DC, Ladha JK (2011) Direct-seeded rice culture in Sri Lanka: lessons from farmers. Field Crop Res 121(1):53–63

Yoshioka Y, Iwata H, Tabata M, Ninomiya S, Ohsawa R (2007) Chalkiness in rice: potential for evaluation with image analysis. Crop Sci 47(5):2113–2120

Chapter 3
A Study on Image Restoration and Analysis

Soumen Kanrar and Srabanti Maji

Abstract Image restoration and analysis are the technique and methodology to recover degraded images through prior knowledge, which includes degradation functions, modeling, and inverse process. This technique can process in two domains. Those domains are spatial domain as well as frequency domain. The various noise models and techniques are improved using MATLAB for the filtering methodology in spatial and frequency domains. We can analyze the different restoration methods given in image enhancement with the help of mean squarer error (MSE) estimation as well as the estimation of signal-to-noise ratio (SNR). Goals for image restoration to determine the types of noise analyze the techniques and determine how the restoration works. Image restoration methodology is distinct with respect to the methodology of image enhancement. Image enhancement uses to emphasize the basic features of an image which enhances an image more presentable for observation. It is not always possible to develop factual data in the prospect of scientific observation. Image enhancement technology is generally provided by imaging packages which are used without an a-priori model to process and produce the image. The book chapter presents the adequate fundamental concepts to enhance the noise reduction technique and resolution loss for image processing.

Keywords Noise reduction · Image restorations · Probability density function · Image enhancement · Filter

3.1 Introduction

The well-known popular part of an objective process is image restoration. It is used to retrieve an image quality by considering some of the prior information based

S. Kanrar (✉)
Department of Computer Science and Engineering, Amity University Jharkhand, Ranchi, India
e-mail: skanrar@rnc.amity.edu

Department of Computer Science, Vidyasagar University West Bengal, Midnapore, India

S. Maji
Department of Computer Science and Engineering, DIT University, Dehradun, India

© The Author(s), under exclusive license to Springer Nature Singapore Pte Ltd. 2022
N. Kumar et al. (eds.), *Advance Concepts of Image Processing and Pattern Recognition*, Transactions on Computer Systems and Networks, https://doi.org/10.1007/978-981-16-9324-3_3

on the degradation phenomenon. This common approach utilized the criterion of goodness, which provides the optimal estimates of results. For the enhancement of applied techniques, researchers are used heuristic concepts to control an image at the psychophysical aspects of the human visual system. It is inclined toward the modeling of degradation and implementing the inverse process to get an initial form of the image. Many techniques are developed and implemented in the spatial domain. Spatial processing is generally used for the deterioration of supplement noise only. Another cases when degradations such as images become a blur, frequency domain filters are the suitable choice. Image enhancement involves multi-focus image fusion which is analyzed with the support of content-adaptive blurring (CAB) to determine image's range and blurring (Farid et al. 2019). It is also involved in smart phones as a part of pattern reorganization and human recognizing with the help of histogram, filters, and algorithm which is related to image restoration (Rashmila and Reshna 2017). Enhancement of image in low contrast is possible with the help of discrete wavelet transformation in some harsh conditions which cause blurring such as underwater (Chaubey and Atre 2017). We have found a blurred image which is analyzed as we can check the histogram of a blurred image to see blurriness. Unsharp masking is also used to improve image's sharpness which plays an important part in image enhancement (Ye and Ma 2017). A real-life blurred and noisy image is proved to be difficult to restore unless we implement a new technique by estimating the point spread function (PSF) of an image as it is important to restore an image especially in large sizes (Tang 2002). The image quality can be accessed with the help of image matrix to enhance the image with the help of a histogram to produce better results (Arora and Garg 2016). The most interesting part of image enhancement is pattern recognition as there are various methods and it involves such as blur modeling, discrete cosine transform, and automatic correction of an image as you saw in the latest systems such as processors, smartphones (Boscaro et al. 2017). One of the best methods throughout image restoration technique is the blind deconvolution. It is most challenging in a case of motion deblurring as basic deblurring methods are not enough to restore the image of an object moving particles such as race car (Chen et al. 2011). The speed of deblurring an image depends on the size of an image. For example, small or medium size of image which has low-medium pixels can be deblurred quickly while the large sizes have large pixel that takes few seconds (Cho and Lee 2009). First, an image is corrupted with blurring and noise when we take a picture of an object which we are interested in. Removing blur and noise is mandatory in order to get a clear picture of an object we desired, and it also depends on an atmosphere of an object (Dash et al. 2011). A camera movement is the main cause of getting a motion-blurred image like our hands get shaken produced a blurred image or a bad quality camera shows a blurred and noisy image. With the help of the Lucy–Richardson algorithm and Weiner filter, the image we desired will be restored easily as we need to use methods with the help of given algorithms of given methods, respectively (Dobes and Machala 2007). A regularized function is useful in non-blind deconvolution technique to produce a better result as we compare with regularized function with blind deconvolution (Javaran et al. 2017). We also compare Lucy–Richardson algorithm and Weiner filter with these methods as well to analyze the methods of image restoration. With the help of blind deconvolution, we can restore an image by removing the blur of an image which is very useful in pattern reorganization and object identification (Kanthan and Sujatha 2016). In modern and latest times, various methods of image enhancement are used for face recognition to protect the user's account and give a clear texture of an image of a user who accesses his account by the use of a video camera (Kapil and Abhilasha 2015). Depending on the point spread function (PSF), parameters of PSF effective during deblurring of motion image (Kerouh et al. 2017). The PSF is important as it is used for motion deblurring and recognizes the pattern well for image processing (Lee and Qiu 2016). Blur kernel is also useful to restore an image in low-rank images (Lokhande et al. 2006). A motion blur is an interesting cue in image enhancement as we can see a blur of an animated image when we take a picture of a moving object. A motion blue is somehow useful to analyze the image deeply (Ma et al. 2017). With the help of image smoothing methods including median filter, mean filter, adaptive filter, etc., we can check the performance of an image deblurring process (Navarro et al. 2011). Due to handshaking, object moving, incorrect focus, etc., an image will be shown as blurred and inaccurate and quality of an image is decreased. It is not easy to take a picture of an object accurately (Surya Prabha and Satheesh Kumar 2016). A quantization of an image with the help of Fourier transform is used to robust the blurred image and improve its quality of an image whether a motion one or inaccurate one as the quantization will work (Ragab et al. 2016). A blind convolution is useful to restore the image's quality from blurred image even a camera shake appears. Then a method gives us a clear image so we can have a cleared motion image (Rahtu et al. 2012). Recently, some new technologies are developed to enhance an image automatically like some latest cameras and computerized software which is useful to get a clear image even a camera shakes (Shao-Jie 2009). Also, an image enhancement helps us for comparison between a blurred and noisy image with a restored image as we can determine its PSF individually (Tashiro 2017). We can regularize a motion-blurred image of a particular object as vector per pixel, and with the help of regularized filter, we can see a clearer motion image as deblurred (Tico and Pulli 1521). Image depth is also described in order to determine a sequence of blurring and deblurring an image, we can observe its depth like sharpness and contrast (Wang et al. 2007). Contrast enhancement is very useful to calculate its histogram to observe the effect of an image (Tung and Hwang 2017). An image enhancement is also used to convert an image of low resolution to high resolution. For example, when we see a video from YouTube, there are many resolutions from low to high like 144, 240, 360, 480, 720, and 1080 p (Wang and Chen 2017). A histogram of an image is useful to determine its quality and analyze it as we mostly need a high-resolution image in greater quality as most people need a good quality image (Xiao et al. 2016). With the help of point spread function (PSF), we can analyze the image's frequency domain and motion blur in order to produce a clear motion image (Xu et al. 2016). Hence, it is not easy to develop a motion image quickly because it takes time to process the methods to enhance an image properly (Yu et al. 2016). Image splicing is also involved to merge the object and background from different atmospheres; however, the analysis and proper method for deblurring an image are also necessary (Zahi and Yue 2014). Digital image reconstruction involves scaling, image interpretation, geometric correction, and revolution are utilized with the help of computerized image recreation, and we can produce different images in different sizes (Liu et al. 2013). Image fusion can be performed with the help of wavelet transform which is mostly used for medical purposes (Ramesh et al. 2017). A grayscale image can be deblurred as various methods are used; however, Weiner filter and Fourier inverse filter are mostly used alongside Lucy–Richardson Algorithm rather than blind deconvolution, median filter, mean filter, and other types of filters. Normally, an image restoration compares an original image and deblurred image by the use of histogram analysis (Sale and Sawant 2016). Mostly, the captured images possess a number of degradations, for example, a blur, or uneven illumination, or low resolution. That is why image restoration is needed to remove the degradations to get a clear image for important purposes (Sankhe et al. 2011). Color images are indeed special as even if it is blurred, restoration of color images to high quality is important in modern days (Sharma and Sharma 2016). An image blur needs to be analyzed and observed to see how the motion and shaking affect the image greatly. This is why mapping of image blur is needed alongside PSF (Roubeki 2009). Even a high-quality capture motion image that has degradations needs some proper methods to quantize an image and remove the degradations in order to deblur a high-quality motion image (Su 2011; Kanrar et al. 2012, 2017, 2016, 2021). The resolution of captured images can enhance through the binarization technique (Mukherjee and Kanrar 2010). In racing games, motion blur is mostly involved to make it realistic and it will remain unaffected to the player's experience when gaming (Shan et al. 2008). However, a motion blur makes a document image into low quality as restoration is needed to convert the image's quality from low to high as an improvement with the help of various parameters (Sharan et al. 2013). Motion deblurring is also described for optimization of a particular image as a motion blur occurs mostly when we take pictures or videos from cameras. Commonly modern technologies have new methods to give a clear image by motion deblurring which works automatically with the help of little effort (Su 2012).

3.2 Image Restorations and Analysis

An image enhancement and restoration is mostly used for medical purposes, scientific purposes, photography, photoshop, military, and civilian purposes. As we had noticed in various papers, motion blur is commonly involved during photography and photoshop as well. We need to perform some techniques to deblurring and restore the motion image and provides us a high-quality image. We take several photographs for ourselves, and the quality of an image is not always dependent on cameras. That is why the motivation for image analysis is necessary for personal stuff. For medical purposes, an image must be good in a quality, perfect combination of edge, and contrast to give a proper result of an image to determine the condition of a patient by doctors and scientists. A frequency domain of an image will motivate us to analyze its quality by adding blur and noise and then restore it by using restoration techniques. As we have noticed, various images are captured and saved by many photographers; however, losing concentration such as defocus, handshaking, and moving objects causes an image to get blurred and noisy if pollution takes place. Detection of blur and noise is needed as a photographer must focus on images to check the problem and then restore it by the use of deblurring with the help of modern technologies. Each number of pixel of an image is very important to analyze it. For military and civilian purposes, each image must be taken and with the help of restoration techniques which are available in the latest gadgets like the latest cameras, smartphones, an image will be shown clearly before we take action. Many people demand high-quality images this is why image restoration is needed as it is important to provide better images.

The main goal in this chapter is to analyze different image restoration methods and then perform an improved image fusion to compare which method is best for image fusion. We can perform a methodology to analyze the image's pixel, edge, and various formats. Deblurring methods are useful for many reasons to develop a proper image. We understand that it is difficult to develop a high-quality image, it required enhancement in methodology and implements with the help of a digital camera only. The most interesting fact is when we play a racing game; we take a screenshot during playing. Blurring also takes place during screenshots as not always screenshots take pictures accurately. This is why restoration is used as we can focus to improve the screenshot for ourselves. The techniques of restoration are blind deconvolution, Weiner filter, Lucy–Richardson algorithm, and regularized filter. We can implement a blurred and noisy image and restore with each method and then compare to produce the different results.

3.2.1 Basic Requirement for Image Restoration and Analysis

As displayed in Fig. 3.1, degradation method defines with a supplement noise term. It is implemented on an input image (x, y). Here, we consider as (x, y) to generate a contaminated image g (x, y). Let us consider $g (x, y)$, possess basic information about the degradation function. Let say (H), with some of the information about

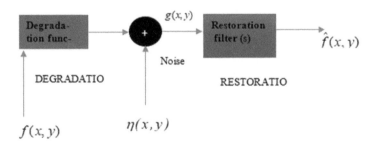

Fig. 3.1 Image degradation model

supplements noise term say, (x, y). The object restoration obtains through an estimate function $\hat{f} x, y)$ for the initial image.

A researcher's basic goal, estimate output is closer to the initial input image. In a common view, the more we get information insight about (H) and η, we proceed near to $\hat{f} x, y)$. Now, we could easily estimate (x, y).

In the spatial domain, contaminated image denoted as:

$$g(x, y) = h(x, y) * f(x, y) + \eta(x, y) \tag{3.1}$$

Here, $h(x, y)$ represents the spatial of the degradation function. The symbol '*' represents convolution.

The degraded image in the frequency domain is expressed as:

$$G(u, v) = H(u, v) F(u, v) + N(u, v) \tag{3.2}$$

where the capital letter's term represents the Fourier transforms. Those are used in the previous Eqs. (3.1) and (3.2).

3.2.2 Noise Models in Image

The origin of noise inside the digital images is spread through image obtainment. Its achievement is influenced through various circumstances, like environmental conditions and perceived parts. Obtaining images with a charge-coupled device (CCD) camera here light levels and temperature sensor are main parts which influence the noise in obtaining images. However, the images can have distorted due to the existence of channel interference during the carry.

3.2.3 Spatial and Frequency Properties

The frequency content of noise in the Fournier sense represents the frequency properties (i.e., in contrast to frequencies of an electromagnetic spectrum). Generally, white noise is the constant of the Fourier spectrum of noise.

This vocabulary is the oddment from material properties of white light which accommodate nearby full frequencies in the noticeable spectrum with alike quantity. It is not so hard to present that the Fourier spectrum of a function contains nearly possible frequencies with alike quantity is a constant. Except for structural periodic noise, we generally consider, noise is independent of structural coordinates. The pixel values and values of noise components in images are independent.

However, this conjecture is moderately inoperative. We have found few applications in the arena of quantum-limited imaging. Those are exhibited in nuclear-medicine imaging as well as in X-ray.

The complexities with the spatially dependent and correlated noise require further exploration.

3.3   Operation of Probability Density Function

When we consider the noise component model, it requires a spatial noise descriptor for the study of statistical exploit of severity values. General probability density function (PDF) used in image processing describes as follows.

3.3.1   Gaussian Noise Model

This noise model is presented by using the probability density function of the normal distribution form. It mostly acquires the values from the Gaussian distribution function.

The PDF can be depicted by

$$p(z) = \frac{1}{\sqrt{2\pi}\sigma} e^{\frac{-(z-w)^2}{2\sigma^2}} \quad -\infty \leq z \leq \infty \tag{3.3}$$

Fig. 3.2 Probability density function of Gaussian noise model

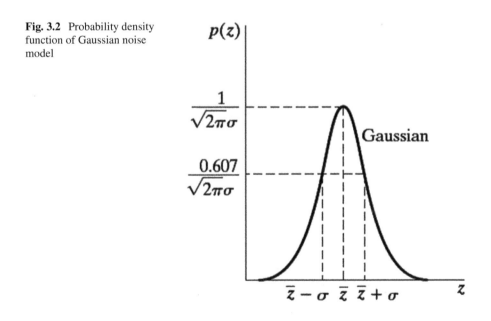

Fig. 3.3 Probability density function of Rayleigh noise model

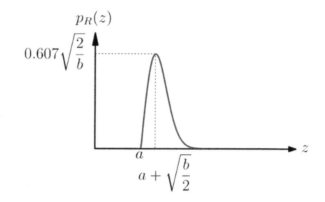

Here, z is denoted as intensity. Here, w is the average value of z, with its standard deviation. The standard deviation squared represents a variance of z. Now Fig. 3.2 exhibits a draw for Eq. (3.3).

3.3.2 Rayleigh Noise Model

The PDF function of Rayleigh noise model can be represented by

$$P(z) = \begin{cases} \frac{2}{b}(z-a)e^{-((z-a)^2)/b} \\ 0 \end{cases} \text{for} \begin{cases} z > a \\ z \le a \end{cases} \tag{3.4}$$

The mean is given by

$$\mu = \frac{a+b}{2} \tag{3.5}$$

And variance is given by

$$\sigma^2 = \frac{(b-a)^2}{12} \tag{3.6}$$

A plot of Eqs. (3.4), (3.5), and (3.6) is shown in Fig. 3.3.

3.3.3 Erlang (Gamma) Noise Model

The PDF of gamma noise is given by

Fig. 3.4 Probability density function of Erlang (Gamma) noise

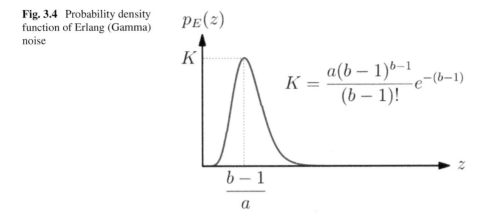

$$P(z) = \begin{cases} \frac{a^b z^{b-1}}{(b-1)!} e^{-az} & \text{for } z \geq 0 \\ 0 & \text{for } z < 0 \end{cases} \tag{3.7}$$

The mean is given by

$$\mu = \frac{b}{a} \tag{3.8}$$

And variance by

$$\sigma^2 = \frac{b}{a^2} \tag{3.9}$$

A plot of Eqs. (3.7), (3.8), and (3.9) is shown in Fig. 3.4.

3.3.4 Exponential Noise Distribution

The probability density function of an exponential noise model is presented by

$$P(z) = \begin{cases} ae^{-az} & \text{for } z \geq 0 \\ 0 & \text{for } z < 0 \end{cases} \tag{3.10}$$

Here, $a > 0$.
The mean is given by

$$\mu = \frac{1}{a} \tag{3.11}$$

And variance by

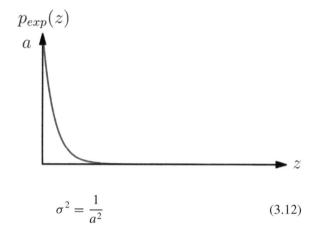

Fig. 3.5 Probability density function of exponential noise model

$$\sigma^2 = \frac{1}{a^2} \tag{3.12}$$

A plot of the above equations is presented in Fig. 3.5.

Uniform Noise Model: One of the commonly used image noise models is a uniform noise model. Here, the noise considers the values from a closed interval $[a, b]$. Values are uniformly distributed. Probability distribution function is presented by

$$P(z) = \begin{cases} \frac{1}{b-a} & \text{if } a \leq z \leq b \\ 0 & \text{otherwise} \end{cases} \tag{3.13}$$

The mean is given by

$$\mu = \frac{a+b}{2} \tag{3.14}$$

And variance by

$$\sigma^2 = \frac{(b-a)^2}{2} \tag{3.15}$$

A plot of Eqs. (3.13), (3.14), and (3.15) is shown in Fig. 3.6.

3.3.5 Impulse (Salt and Pepper) Noise Model

The probability distribution of the impulse noise is presented by

$$P(z) = \begin{cases} P_a & \text{for} & z = a \\ P_b & \text{for} & z = b \\ 0 & \text{otherwise} \end{cases} \tag{3.16}$$

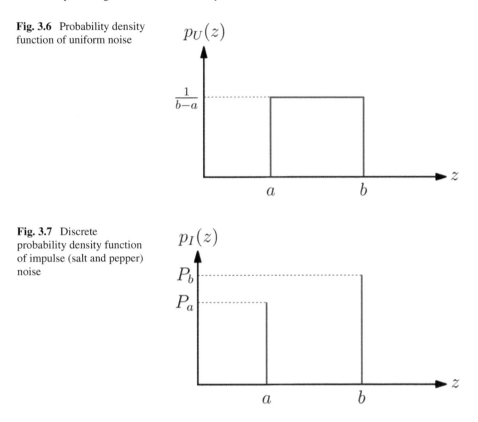

Fig. 3.6 Probability density function of uniform noise

Fig. 3.7 Discrete probability density function of impulse (salt and pepper) noise

Intensity value (*b*) greater than (*a*) then it seems to be a spotlight, as a dot inside the image. The reverse level will appear like a spot dark dot.

Whether or are zero values, then the impulse noise will be unipolar. A plot of Eq. (3.16) is presented in Fig. 3.7.

3.4   Problems in Image Enhancement

There are certain problems in image enhancement such as:

1. When we take a picture of an object, whether still or motion one; defocus and inaccuracy occur as an image is shown as blurred.
2. Noise occurs if a camera does not work properly, or pollution occurs such as air pollution to reduce clarity and gives a low-quality image.
3. Blurring also takes place in screenshots when we play a game and wanted to show that to some friends. For example, when we play a racing game, a motion blur occurs in the background of a game while a car is shown clearly or as a blurred image.

4. Camera shake is one of the main problems which cause image degradation, and all focus for an object is lost. This reduces the quality of photographs greatly.
5. Taking photographs from a far distance can show us a blurred image as it is not easy to take a proper picture from a long distance as a closed distance is easier.
6. A motion image is not easy to capture with accuracy. It will have blurring as an object moves in motion.
7. An image is considered as out of focus blurred when an object does not go along with the camera lens.
8. The demand for high-quality images is increasing because such images are used by many people including military, civilian, medical, exhibition, etc.
9. In very low light conditions such as a night, we cannot see an object properly and an image is produced in dark contrast. Thus, low light conditions have more noise than other conditions.
10. Due to bad weather such as fog, rain, etc., the image is produced as blur and noisy because the camera lens gets corrupted if it gets affected by rainwater, fog, etc.

3.5   Methods Used in Image Restoration

3.5.1   Inverse Filtering

Direct inverse filtering is one of the commonly known simplest approaches for image restoration. In this method, the Fourier transforms of the image $f(u, v)$ are computed by dividing the Fourier transform. A degraded image by the Fourier transforms is used for an estimate. Now, degradation function is expressed as

$$f(u, v) = \frac{G(u, v)}{H(u, v)} \tag{3.17}$$

This procedure works effectively when the convention is no supplement noise present in the degraded image. The demeaned image expresses with following Eq. (3.18).

$$g(x, y) = f(x, y) * h(x, y) \tag{3.18}$$

However, the noise gradually included and becomes a degraded image. The resultant effect is that the direct inverse filtering becomes very poor. By, substituting $G(u, v)$ proceeded to Eq. (3.19).

$$\hat{f}(u, v) = f(u, v) + \frac{N(u, v)}{H(u, v)} \tag{3.19}$$

Equation (3.19) represents the direct inverse filtering fails. It presents the supplement noise induced into inside a degraded image. The noise is random. Due to that, it is very challenging to discover noise spectrum $N(u, v)$. Deconvolution mechanism uses Lucy–Richardson algorithm (DLR). This mechanism is a non-blind routine for image restoration. It is used to makeover a discomfited image. It is mostly blurred by a recognized point spread function (PSF).

3.5.2 Weiner Filter

Weiner filtering is made on the concept of non-blind methodology. It is designed to maintain the reconstructing of a degraded image with a known point spread function. After completed previous stage that reverses blurring together. It is performed the deconvolution through the high-pass inverse filter. It also removes the noise with the help of a density operation, i.e., low-pass filter. It compares and normal estimation of any preferred noise-free image and induces degraded image information to Wiener filter generally distorts by supplement noise. Output image is obtained using a filter with expression (3.20).

Here, f is original input image, and n represents noise. Now, $\hat{f}$ is estimated output image and g is response from Wiener filters.

$$\hat{f} = g + (f + n). \tag{3.20}$$

Weiner filter is mostly useful in order to analyze each image's working and also gives us a clearer picture alongside histogram effects. In terms of mean square error, Wiener filtering behaves as most optimal. The process of noise leveling and antithetical filtering gradually minimizes overall mean square error. Wiener filtering is a linear estimator for an original image.

3.5.3 Algorithm for Using Weiner Filter

Step 1: Read Image in any format. (JPG, PNG, JPEG, GIF, etc.).
Step 2: Adding Blur in the given image. (Depends on Len, Theta, PSF, Type).
Step 3: Using Deconvwnr, Restore the Blurred Image.
Step 4: Add Blur and Noise to the Image (Depends on Noise Mean, Noise Variation, and Type).
Step 5: Restore the Blur–Noisy Image according to NSR—Noise Scale Ratio Image (Using Deconvwnr). (Step 6 to Step 8).
Step 6: NSR $= 0$, Non-Quantized Image
Step 7: NSR $= 0$, Quantized Image.
Step 8: Quantized Image, Estimated NSR.
Step 9: Terminate the program.

Time Complexity of Weiner Filter Algorithm is $O\left(n^2 \log n\right)$.

3.5.4 Lucy–Richardson Algorithm

L.B. Lucy and W. Richardson have invented this concept. Deconvolution mechanism uses regularized filter as a base for non-blind image restoration. It utilizes to restore a degraded image to become blurred by a pre-known point spread function In this iterative procedure, where the pixels belong to viewed images are used point spread function. A latent image is presented through the below expression.

$$d_i = \sum p_{ij} u_j \qquad (3.21)$$

Here, d_i examines value at the pixel location (i). And p_{ij} is point spread function. Fraction of light is impending from accurate position j. Actually, it observes at point location (i). Latent image pixel value at the position 'j' inside an image denotes u_j. Objective of this method is to calculate the mostly occurred 'u_j' in the presence of an examine value d_j. Now, the point spread function (PSF) p_{ij} is presented in Eq. (3.22) as follows:

$$u_j^{t+1} = u_j^t \sum_{i \frac{d_i}{c_i} p_{ij}} c_i = \sum_{j p_{ij} u_j} (t) \qquad (3.22)$$

3.5.5 Lucy–Richardson Algorithm

Step 1: Read Image in any format. (JPG, PNG, JPEG, GIF, etc.).
Step 2: Add Blur and Noise to the Image (Depends on Noise Mean, Noise Variation, and Type).
Step 3: Restore Image according to NUMIT—Number of Iterations. Example: NUMIT = 5, 15, etc.
Step 4: Control the Noise Amplification with the help of Damping after Step 3.
Step 5: Create a Sample Data.
Step 6: Observe The Data by Simulating its Blur, Providing a Weight Array, PSF, etc.
Step 7: Restore The Data as Normal, Binned, PSF, etc.
Step 8: Terminate the Process.

Time Complexity of Lucy–Richardson Algorithm is $O\left(n^2\right)$.

3.5.6   Regularized Filter Used in Image Restoration

Regulated filter follows the deblurring method. It is used to deblurred an image by considering the deconvolution function to reconverge. It is very effective when we know limited information about additive noise. A regularized filter always focuses on the accuracy of an image. Reduce noises are obtained only. It is possible that blur the edges of an input image with the help of PSF can reduce, and the noises are amplified.

3.5.7   Algorithm for Regularized Filter Implement on Sample Image

Step 1: Read Image in any format. (JPG, PNG, JPEG, GIF, etc.).
Step 2: Add Blur and Noise to the Image (Depends on Noise Mean, Noise Variation, and Type).
Step 3: Restore the Blurred Image with respect to NP – Noise Power.
Step 4: Compare the NP according to normal, large, and small.
Step 5: Apply Edge taper Effect.
Step 6: Restore The Image by the use of LAGRA—Lagrange Multiplier.
Step 7: Compare the LAGRA according to normal, large, and small.
Step 8: Restore The Image by Constrain it with 1-D Laplacian.
Step 9: Terminate the Process.

Time Complexity of Regularized Filter Algorithm is $O\left(N^2 \log N\right)$.

3.6   Blind Deconvolution for Blur Image

The blind deconvolution mechanism allows for recuperation from a target scene. It is from a set of "unclear" images in point spread function. In blind deconvolution, point spread function is evaluated from a collection of images. Deconvolution is also performed. Blind deconvolution is utilized in the field of medical imaging and astronomical images. It can be utilized interactively. Each iteration is an improved estimation of the point spread function. One of the applications for a non-interactive algorithm is based on exterior information. It explores point spread function values that better estimate point spread function PSF. It is useful for quicker convergence.

3.6.1   Blind Deconvolution Algorithms for Sample Image

Step 1: Read Image in any format. (JPG, PNG, JPEG, GIF, etc.).
Step 2: Add Blur and Noise to the Image (Depends on Noise Mean, Noise Variation, and Type).
Step 3: Restore the Blurred Image according to PSF – Point Spread Function.
Step 4: Compare Undersized and Oversized PSF.
Step 5: Restore the Image by the use of INITPST—Initial Estimate of PSF.
Step 6: Terminate the Process.

Time Complexity of Blind Deconvolution Algorithm is $O(N^2 \log N)$.

3.6.2   Methodologies Implemented on Blur and Noisy Sample Image

A novel image fusion method based on discrete wavelet transformation DWT is used. This chapter exhibits a comparison of different image restoration methods. It exhibits to improve the performance of images fusion in presence of unknown image degradation. A methodology is used in the task of image restoration, which will maintain the required information from both images. This chapter provides a general overview of the proposed technique.

3.6.3   Wavelet Transformation in 2-D

Planer scaling functions like $W(x, y)$, and other wavelet functions $\psi^H(x, y)$, $\psi^V(x, y)$, $\psi^D(x, y)$ are crucial foundations for wavelet transforms (Rashmila and Reshna 2017). It is a composition of one-dimensional scaling function φ and the respective wavelet ψ. Those are presented in the following equations (Fig. 3.8).

$$W(y, x) = W(y) * W(x) \tag{3.23}$$

$$\psi^H(y, x) = (y) * W(x) \tag{3.24}$$

$$\psi^V(y, x) = W(y) * \psi(x) \tag{3.25}$$

$$\psi^D(y, x) = (y) * \psi(x) \tag{3.26}$$

Figure 3.9 exhibits, ψ^H the measures of horizontal variations with the corresponding vertical variation ψ^V and detected variation ψ^D along the diagonal

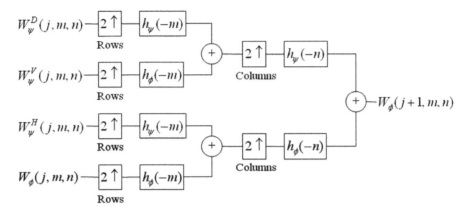

Fig. 3.8 Filter bank of two-dimensional (fast wavelet transform)

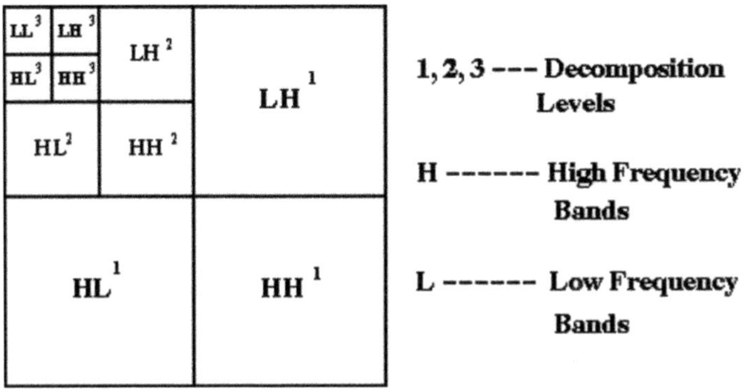

LL³	LH³	LH²		LH¹	
HL³	HH³				
HL²		HH²			
HL¹			HH¹		

1, 2, 3 --- **Decomposition Levels**

H ------ **High Frequency Bands**

L ------ **Low Frequency Bands**

Fig. 3.9 A two-level decomposition of the two-dimensional DWT

Algorithm Discrete Wavelet Transformation.

Step 1: Applying DWT to break down input image into sub-bands.
Step 2: **Compute**: Wavelet coefficient value σ_x and σ_y. (Standard deviation).
Step 3: **Calculate:** $D = \text{abs}(\sigma_x) - \text{abs}(\sigma_y)$.
Step 4: **Calculate:** fused coefficients

$$z = W_X \sigma_x + W_Y \sigma_y$$

Step 5: if $(D < 0)$ then,

$$W_X = 0 \text{ and } W_Y = 1 - W_X$$

Step 6: if $(D > = 0)$ then,

$$W_X = 1 \text{ and } W_Y = 1 - W_X$$

Step 7: **Calculate**: Average coefficients in low-pass residuals.
Step 8: **Reconstruct**: Fused image from the processed sub-bands and the low-pass residual by applying inverse DWT.
Step 9: Terminate the process.
Step 10: **End**.

Time Complexity of DWT algorithm is $O(N^2 \log_2 N)$.

3.6.4 Estimation Method of Signal-to-Noise Ratio (SNR)

Restoration techniques through signal-to-noise ratio (SNR) can be able to scrutinize the difference. The difference between the original as well as restored image is observed.

Wiener filtering restoration algorithm estimated signal-to-noise ratios (SNR) is selected from regulated parameters in Wiener filtering restoration. It is usually defined as,

$$SNR = 10 \log_{10}(\delta_x^2 / \delta_n^2) \tag{3.27}$$

where

δ_x^2 denotes variance of the blurred image.
δ_n^2 senotes variance of noise.

In an image, provincial variance between flat region and edge is distinct from each other.

This region places with large provincial variance. It is relatively flat as exhibited. There is much detailed information present in the places with small local variance.

If the observed quality of an image is good, then it is equitable to consider a region with maximum local variance. Here, least possible regional variance considers as flat sectional. Here (δ_x^2) is considered as the maximum regional variance of the image. Also, (δ_n^2) represents an approximate minimum local variance of the noise. Let consider (y) is the image and the observed local variance in the position (i, j). The expression is presented below equation.

$$\delta_{yt}^2(i, j) = \frac{1}{(2p + 1)(2q + 1)} \sum_{k=-p}^{p} \sum_{l=-q}^{q} [y(i + k, j + l) - \mu_y(i, j)]^2 \tag{3.28}$$

Now,
Variables p and q sizes of a territorial area.
μ_y Territorial mean value

$$\mu_y = \frac{1}{(2p+1)(2q+1)} \sum_{k=-p}^{p} \sum_{l=-q}^{q} y(i+k, j+l) \tag{3.29}$$

The local variance is obtained with the value $p = q = 2$, and the corresponding arrangement is expressed as:

$$c = \frac{1}{(2p+1)(2q+1)} \begin{bmatrix} 1\ 1\ 1\ 1\ 1 \\ 1\ 1\ 1\ 1\ 1 \\ 1\ 1\ 1\ 1\ 1 \\ 1\ 1\ 1\ 1\ 1 \\ 1\ 1\ 1\ 1\ 1 \end{bmatrix} \tag{3.30}$$

So, territorial mean value

$$\mu_y = \frac{1}{(2p+1)(2q+1)} \sum_{k=-p}^{p} \sum_{l=-q}^{q} y(i+k, j+l) \tag{3.31}$$

$$= \sum_{k=-p}^{p} \sum_{l=-q}^{q} y(i+k, j+l).c(k, l) = y * c \tag{3.32}$$

Here the notation '*' denotes the convolution operation. The local variance is rewritten as:

$$\delta_{yt}^2(i, j) = \frac{1}{(2p+1)(2q+1)} \sum_{k=-p}^{p} \sum_{l=-q}^{q} [y(i+k, j+l) - \mu_y(i, j)]^2 \tag{3.33}$$

$$= \sum_{k=-p}^{p} \sum_{l=-q}^{q} \{[y(i+k, j+l) - \mu_y(i, j)]^2.c[k, l]\} \tag{3.34}$$

$$= (y - \mu_y)^2 * c \tag{3.35}$$

Now, SNR of an image says (y) is likely as a ratio of topmost regional variance with the smallest topical variance.

$$\text{SNR} = 10\log_{10}\left(\max(\delta_{yL}^2)/\min(\delta_{yL}^2)\right) \tag{3.36}$$

3.6.5 Estimation of the Gaussian Point Spread Function (PSF)

Gaussian 'point spread function' uses in many optical computation and imaging systems. It is expressed as follows:

$$h(m, n) = \begin{cases} \frac{1}{\sqrt{2\pi}\sigma} \exp\{\frac{-1}{2\sigma^2}(m^2 + n^2)\} & (m, n) \int R \\ 0 \end{cases} \qquad (3.37)$$

Here, we consider standard deviation is σ with a supporting region R.

R represents a matrix with a size of $(K \times K)$. Now, K considers an odd integer. The parameters, namely size (K) and standard deviation (σ), are used to identify Gaussian point spread mathematical expressions. Multiple curves are generated for different sizes and values of the Gaussian 'point spread function.' The real size and standard deviation are approximated through analyzing a relationship between obtained curves. Basic criteria to estimate the parameter's asset value of Gaussian 'point spread function' is summarized as follows. Estimated size is approximated through distance between curves decreases evidently. The estimated standard deviation is assumed where the corresponding curve increases. With the help of two thresholds (say) TH1 and TH2, we can have approximate parameters of Gaussian 'point spread function.'

Let consider 'e' is an evaluated error. When a gap between curves is smaller than a predefined value TH1, it produces measured size $(\hat{K})$ for Gaussian 'point spread function.' Standard deviation 'e' is the distance measure as a pure variation between cycle numbers (j). The slop estimation errors with respect to various standard deviations on an approximated curve are obtained. If measure deviation of the slop is greater than the predefined threshold value TH2, then it is the estimated deviation $(\hat{\sigma})$.

3.6.6 Comparison of Mean Square Error (MSE)

The differences in pixels are measured with the help of mean square error.

It is used to calculate the major deviation between the recovered images from the original image.

Obtaining a lesser value represents better restoration. It can be clearly seen by observing a graph.

The value of MSE for FFT is higher. However, this value is lower for the case of DWT Haar transforms. We can compare to other wavelet families, for example, Symlet, Daubechies, and Coiflet.

The mean square error is expressed by:

$$e^2 = E\left\{ \left(f - \hat{f} \right)^2 \right\}$$ (3.38)

The notation $E\,()$ denotes expectation of argument.

We consider the noise, and images are uncorrelated. Intensity levels belong to the estimate. It is generally a linear function based on levels in a degraded image.

$$\hat{F}(u, v) = \left[\frac{H^*(u, v)S_f(u, v)}{\left(S_f(u, v)|H(u, v)|^2 + S_\eta(u, v) \right)} \right] G(u, v)$$ (3.39)

$$= \left[\frac{H^*(u, v)}{\left(|H(u, v)|^2 + \frac{S_\eta(u,v)}{S_f(u,v)} \right)} \right] G(u, v)$$ (3.40)

$$= \frac{1}{H(u, v)} \left[\frac{|H(u, v)|^2}{\left(|H(u, v)|^2 + \frac{S_\eta(u,v)}{S_f(u,v)} \right)} \right] G(u, v)$$ (3.41)

Output of a complex quantity with respect to its conjugate is equivalent to magnitude of a squared of complex quantity.

The terms used in the above equation are described as below.

The degraded function is presented by $H(u, v)$.

where the complex conjugate of $H(u, v)$ is presented by $H^*(u, v)$ and expressed as

$$|H(u, v)|^2 = H^*(u, v)H(u, v)$$

The power spectrum of the noise is represented by $S_\eta(u, v) = |N(u, v)|^2$.

The power spectrum of the undergraded image is represented by the expression

$$S_f(u, v) = |F(u, v)|^2$$

The mean square error is expressed in a statistical form with approximation to the original and restored images in below expression.

$$\text{MSE} = \frac{1}{MN} \sum_{x=0}^{M-1} \sum_{y=0}^{N-1} \left[f(x, y) - \hat{f}(x, y) \right]^2$$ (3.42)

3.7 Analysis of the Image Restoration Methods

We are observing the results of distinctive methods with the help of different images to analyze the SNR, MSE, and Peak-SNR for Fig. 3.10, as shown in Figs. 3.11, 3.12, and 3.13.

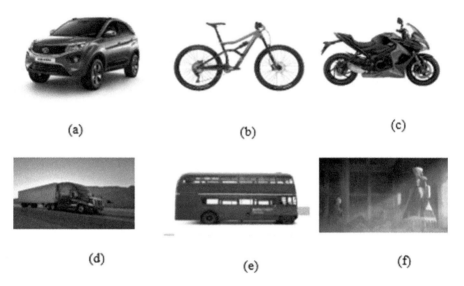

(a)　　　　　　(b)　　　　　　(c)

(d)　　　　　　(e)　　　　　　(f)

Fig. 3.10 Sample images we observed: **a** Car, **b** Cycle, **c** Bike, **d** Truck, **e** Bus, and **f** Master

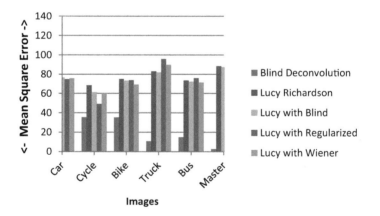

Fig. 3.11 2-D graph shown for MSE comparison given in Table 3.1

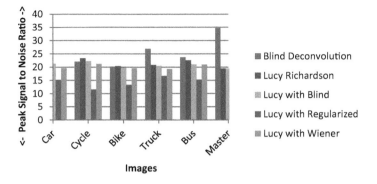

Fig. 3.12 2-D graph shown for Peak-SNR comparison given in Table 3.2

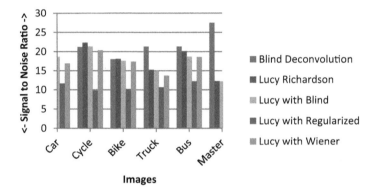

Fig. 3.13 2-D graph shown for SNR comparison given in Table 3.3

3.7.1 Mean Square Error (MSE) Comparison

See Table 3.1.

Table 3.1 Comparison of MSE between images

Images	Blind deconvolution	Lucy–richardson	Lucy with blind	Lucy with regularized	Lucy with wiener
Car	21.5341	75.3876	76.5631	74.8505	75.8233
Cycle	35.5057	68.5989	61.2214	49.1187	60.0907
Bike	35.2850	74.9841	73.3451	73.9043	69.4854
Truck	10.7785	83.0347	82.0596	95.7333	89.7720
Bus	14.8763	73.5944	72.3822	76.1124	71.5879
Master	2.7732	88.2262	87.5551	118.7578	95.1823

Table 3.2 Comparison of Peak-SNR between images

Images	Blind deconvolution	Lucy–richardson	Lucy with blind	Lucy with regularized	Lucy with wiener
Car	22.9335	21.3078	21.2784	15.2395	19.6346
Cycle	22.0635	23.3064	22.2190	11.5833	21.2402
Bike	20.1821	20.3846	19.7861	13.3612	19.5671
Truck	26.8822	20.7981	20.4303	16.6862	19.2895
Bus	23.6936	22.5350	21.0604	15.3043	20.9768
Master	34.7160	19.4317	19.4443	15.5671	18.3624

Table 3.3 Comparison of SNR between images

Images	Blind deconvolution	Lucy–richardson	Lucy with blind	Lucy with regularized	Lucy with wiener
Car	20.2697	18.5619	18.6146	11.7123	16.9323
Cycle	21.1578	22.3018	21.3133	9.8062	20.3340
Bike	17.9914	18.1094	17.5953	10.1900	17.3667
Truck	21.2840	15.2572	14.8322	10.6711	13.7371
Bus	21.2986	20.0446	18.6654	12.1986	18.5822
Master	27.4963	12.2508	12.2246	7.6452	11.2201

3.7.2   Peak-Signal-to-Noise Ratio (Peak-SNR) Comparison

The peak-SNR of various image techniques is analyzed and compared with the help of MATLAB, and the result is produced given in Table 3.2.

3.7.3   Signal-to-Noise Ratio (SNR) Comparison

Signal-to-noise ratio (SNR) is compared between images, and techniques are shown in Table 3.3.

3.8   Conclusion and Feature Work

This demonstrated the various restoration performances that had been utilized to rebuild an original image from a degraded image. To summarize, it is concluded that filters play an important role to construct a good quality image. Now mostly four techniques have been simulated as well as compared. In the results, the performance of the LR Weiner filter is better than the regularized filter in case of all types of noises.

Similarly, the performance of blind deconvolution is better to regularized filter and poor as compared with Lucy–Richardson techniques.

An image restoration is useful and proves as we can improve the motion image by using restoration techniques, and it is scientifically used by experts who can make the clear motion image better, clear, and picture-perfect. With the help of MSE comparison, we can calculate how the techniques work well in MATLAB and we can produce the image in high quality and best motion. This research helped us to analyze image restoration and different techniques. Also, this research is very important to avoid blurring, noise, and pollution of the images we love. It also helps us to get some new experience about image-related activities such as photography and selfie with the help of image restoration and enhancement and also provided more experience about digital image processing.

References

Arora I, Garg NK (2016) Bi-featured image quality assessment with the hierarchical image quality enhancement algorithm. In: Proceeding: ICICT-IEEE, pp 1–6. https://doi.org/10.1109/INVENT IVE.2016.7824834

Boscaro et al (2017) Pattern image enhancement by automatic focus correction. Microelectron Reliab 76–77:249–254. https://doi.org/10.1016/j.microrel.2017.07.012

Chaubey A, Atre A (2017) A hybrid DWT-DCLAHE method for enhancement of low contrast underwater images. In: proceeding ICECA-IEEE, pp 196–201. https://doi.org/10.1109/ICECA. 2017.8203670

Chen J, Xie Z, Sheng B, Ma L (2011) Motion deblurring from a single image using gradient enhancement. In: Proceeding: VRCAI ACM, pp 293–300. https://doi.org/10.1145/2087756.208 7800

Cho S, Lee S (2009) Fast motion deblurring. In: Proceeding: ACM SIGGRAPH Asia, Article No. 145, 2009. https://doi.org/10.1145/1661412.1618491

Dash R, Sa PK, Majhi B (2011) Restoration of images corrupted with blur and impulse noise. In: Proceeding: ICCCS ACM, pp 377–382. https://doi.org/10.1145/1947940.1948019

Dobes M, Machala L (2007) Restoration of motion blurred images. In: Proceeding: SCCG ACM, pp 87–92. https://doi.org/10.1145/2614348.2614361

Farid MS, Mahmood A, Al-Maadeed SA (2019) Multi-focus image fusion using content adaptive blurring. Inf Fus 45:96–112.https://doi.org/10.1016/j.inffus.2018.01.009

Javaran TA, Hassanpour H, Abolghasemi V (2017) Non-blind image deconvolution using a regularization based on re-blurring process. Comput Vis Image Understand 154:16–34.https://doi.org/10.1016/j.cviu.2016.09.013

Kanrar S (2012) Analysis and implementation of the large-scale video-on-demand system. Int J Appl Inf Syst (IJAIS), 1(4). arXiv: 1202.5094. ISSN: 2249-0868

Kanrar S, Mandal NK (2017) Video traffic analytics for large scale surveillance. Multimedia Tools Appl 76(11):13315–13342.https://doi.org/10.1007/s11042-016-3752-0

Kanrar S, Kumar Mandal N (2016) Video traffic flow analysis in distributed system during interactive session, 2016:1–16. Article ID 7829570https://doi.org/10.1155/2016/7829570

Kanrar S, Agrawal S, Sharma A (2021) Vehicle detection and count in the captured stream video using machine learning. Studies in Computational Intelligence book series (SCI, volume 968), pp 79–112. https://doi.org/10.1007/978-981-16-0935-0_5

Kapil D, Abhilasha (2015) Face recognition of blurred images using image enhancement and texture features. NGCT-IEEE, pp 894–897. https://doi.org/10.1109/NGCT.2015.7375248

Kerouh F, Ziou D, Serir A (2017) Histogram modelling-based no reference blur quality measure. Sig Proc Image Commun 60:22–28. https://doi.org/10.1016/j.image.2017.08.014

Lee C-H, Qiu Z-Y (2016) Motion image de-blurring system based on the effectiveness parameters of point spread function. In: Proceeding: ICNCC ACM, pp 27–31. https://doi.org/10.1145/303 3288.3033308

Liu G, Wang J, Lian S, Dai Y (2013) Detect image splicing with artificial blurred boundary. Math Comput Modell 57(11–12):2647–2659. https://doi.org/10.1016/j.mcm.2011.06.026

Lokhande R, Arya KV, Gupta P (2006) Identification of parameters and restoration of motion blurred images. In: Proceeding SAC '06: proceedings of the 2006 ACM symposium on applied computing, pp 301–305. https://doi.org/10.1145/1141277.1141347

Ma T-H, Huang T-Z, Zhao X-L, Lou Y (2017) Image deblurring with an inaccurate blur kernel using a group-based low-rank image prior. Inf Sci 213–233. https://doi.org/10.1016/j.ins.2017.04.049

Mukherjee A, Kanrar S (2010) Enhancement of Image Resolution by Binarization. Int J Comput Appl 10(10):15–19. https://doi.org/10.5120/1519-1942

Navarro F, Castillo S, Seron F, Gutierrez D (2011) Perceptual considerations for motion blur rendering. ACM Trans Appl Percep 81(3):1–15, Article No.: 20. https://doi.org/10.1145/201 0325.2010330

Ragab AN, Rehan MM, Hassan Y (2016) Partially blurred images restoration using adaptive multistage approach. In: Proceeding: CCECE-IEEE, pp 1–5. https://doi.org/10.1109/CCECE.2016.7726830

Rahtu E, Heikkila J, Ojansivu V, Ahonen T (2012) Local phase quantization for blur-insensitive image analysis. Image vis Comput 30(8):501–512. https://doi.org/10.1016/j.imavis.2012.04.001

Ramesh AVK Manikandan M (2017) Enhancement of interpolation mechanism in large and scalable images. In: Proceeding ICSCN-IEEE, pp 1–4. https://doi.org/10.1109/ICSCN.2017.8085683

Ramesh Kanthan M, Naganandini Sujatha S (2016) Blur removal using blind deconvolution and gradient energy. In: Proceeding: ICCIC-IEEE, pp 1–5. https://doi.org/10.1109/ICCIC.2016.791 9556

Rashmila GV, Reshna T (2017) Intelligent photo: a design for photo enhancement and human identification by histogram equalization, enhancing filters and haar-cascades. In: Proceeding: ICICCS IEEE, pp 1014–1017. https://doi.org/10.1109/ICCONS.2017.8250618

Roubeki FS, Flusser J, Cristobal G (2009) Super-resolution and blind deconvolution for rational factors with an application to color images. Comput J 52(1):142–152.https://doi.org/10.1093/comjnl/bxm098

Sale D, Sawant S (2016) Wavelet based selection for fusion of medical images. In: Proceeding: ICCUBEA-IEEE, pp 1–6. https://doi.org/10.1109/ICCUBEA.2016.7860049

Sankhe PD, Patil M, Margaret M (2011) Deblurring of grayscale images using inverse and wiener filter. In: Proceeding: ICWET ACM, pp 145–148. https://doi.org/10.1145/1980022.1980053

Shan Q, Jia J, Agarwala A (2008) High–quality Motion deblurring from a single image. ACM Trans Graph 27(3):73:1–10. https://doi.org/10.1145/1360612.1360672

Shao-Jie S, Qiong W, Guo-Hui L (2009) Blind image deconvolution for single motion-blurred image. ICIS-IEEE, 491–494. https://doi.org/10.1109/ICICISYS.2009.5357629

Sharan L, Han Neo Z, Mitchell K, Hodgins JK (2013) Simulated motion blur does not improve player experience in racing game, MIC. ACM, pp 149–154, https://doi.org/10.1145/2522628.252 2653

Sharma P, Sharma S (2016) Image processing based degraded camera captured document enhancement for improved OCR accuracy. In: Proceeding: cloud system and big data engineering, IEEE, pp 441–444. https://doi.org/10.1109/CONFLUENCE.2016.7508160

Su B, Lu S, Tan CL (2011) Blurred image region detection and classification. In: Proceeding: (MM) multimedia ACM, pp 1397–1400. https://doi.org/10.1145/2072298.2072024

Su B, Lu S, Lim TC (2012) Restoration of motion blurred document images, SAC. ACM 767–770. https://doi.org/10.1145/2245276.2245424

Surya Prabha D, Satheesh Kumar J (2016) Performance analysis of image smoothing methods for low level of distortion. In: Proceeding: ICACA-IEEE, pp 372–376. https://doi.org/10.1109/ICACA.2016.7887983

Tang H (2002) A combined approach to enhancement of unknown blurred and noisy images. In: Proceeding: ISSIPPN-IEEE, pp 780–783. https://doi.org/10.1109/SIPNN.1994.344795

Tashiro K, Fujie T, Ikei Y, Amemiya T, Hirota K, Kitazaki M (2017) TwinCam: omni-directional stereoscopic live viewing cam-era for reducing motion blur during head rotation. SIGGRAPH ACM Emerg Tech 1–2, Article No: 24. https://doi.org/10.1145/3084822.3084831

Tico M, Pulli K (2009) Image enhancement method via blur and noisy image fusion. In: Proceeding: ICIP-IEEE, pp 1521–1524. https://doi.org/10.1109/ICIP.2009.5413626

Tung S-S, Hwang W-L (2017) Multiple depth layers and all-in-focus image generations by blurring and deblurring operations. Pattern Recogn 184–198. https://doi.org/10.1016/j.patcog.2017.03.035

Wang X, Chen L (2017) An effective histogram modification scheme for image contrast enhancement. Sig Proc Image Commun 58:187–198. https://doi.org/10.1016/j.image.2017.07.009

Wang J, Agrawala M, Cohen MF (2007) Soft scissors: an interactive tool for realtime high quality matting. ACM Trans Graph 26(3):1–6. https://doi.org/10.1145/1276377.1276389

Xiao J, Pang G, Zhang Y, Kuang Y, Yan Y, Wang Y (2016) Adaptive shock filter for image super-resolution and enhancement. J Vis Commun Image Represent 40(Part A):168–177. https://doi.org/10.1016/j.jvcir.2016.06.015

Xu Y, Liang K, Xiong Y, Wang H (2016) An analytical optimization model for infrared image enhancement via local context. Infrared Phys Technol 87:143–152. https://doi.org/10.1016/j.infrared.2017.10.002

Ye W, Ma K-K (2017) Blurriness-guided unsharp masking. In: Proceeding: ICIP-IEEE, pp 3770–3774. https://doi.org/10.1109/ICIP.2017.8296987

Yu C-K, Tsai B-C, Hwang Y-T (2016) An efficient motion blurred image restoration scheme based on frequency domain estimation. In: Proceeding: ICCE-TW-IEEE, pp.1–2. https://doi.org/10.1109/ICCE-TW.2016.7520918

Zahi G, Yue S (2014) Reducing motion blurring associated with temporal summation in low light scenes for image quality enhancement. In: Proceeding: MFI-IEEE, pp1–5. https://doi.org/10.1109/MFI.2014.6997725

Chapter 4
Application of Deep Learning and Machine Learning in Pattern Recognition

E. Fantin Irudaya Raj and M. Balaji

Abstract In today's digital environment, the pattern is everywhere. It is present in many aspects of our daily life. Algorithms can be used to detect or mathematically observe a pattern physically. Vector feature values represent a pattern in the digital world. With the advent of artificial intelligence (AI) techniques in the recent era, there are so many machine learning (ML) and deep learning (DL) models that have been developed. ML is the branch of AI, which can perform tasks like data analysis, analytical model building, and classification without being explicitly programmed. DL is the subset of ML in AI. In DL, mostly artificial neural networks (ANNs) are utilized. It has the capability of work based upon unsupervised learning from the data that is unlabeled and unstructured. Using these DL and ML models, extract the meaningful features from the given image or video is known as pattern recognition (PR). PR is used in many engineering applications such as computer vision, natural language processing, character recognition, robotics, speech recognition, and so on. It is also used in many medical image processing applications and telemedicine. The present work discussed the pattern recognition problem and its various stages in detail. In addition to that, the application of deep learning and machine learning in pattern recognition is also explained briefly.

Keywords Artificial intelligence · Machine learning · Pattern recognition · Artificial neural network · Deep learning

4.1 Introduction

Artificial intelligence (AI) has evolved as a realistic technology in recent years, with beneficial applications in a variety of sectors. Most of these technologies are related

---

E. Fantin Irudaya Raj (✉)
Department of Electrical and Electronics Engineering, Dr. Sivanthi Aditanar College of Engineering, Tiruchendur, Tamil Nadu, India

M. Balaji
Department of Electrical and Electronics Engineering, SSN College of Engineering, Chennai, Tamil Nadu, India
e-mail: balajim@ssn.edu.in

© The Author(s), under exclusive license to Springer Nature Singapore Pte Ltd. 2022
N. Kumar et al. (eds.), *Advance Concepts of Image Processing and Pattern Recognition*,
Transactions on Computer Systems and Networks,
https://doi.org/10.1007/978-981-16-9324-3_4

to pattern recognition (PR). The procedure of identifying patterns with the help of deep learning (DL) and machine learning (ML) algorithms is known as PR. It is the process of classifying data using statistical information or prior knowledge derived from patterns and represented from real-time images or videos. The term PR includes a broad range of information processing problems of great practical significance like natural language processing, medical diagnosis, classification of handwritten characters, speech recognition, detecting a fault in the machinery, etc. Usually, these are challenges that many people seem to overcome effortlessly. However, in many circumstances, their computer-based solution has proven to be quite challenging. Therefore, it is vital to take a proactive approach in order to have the best chance of finding realistic solutions. A statistical framework is the most natural and general framework for formulating solutions to PR challenges.

Raw data is translated and processed into a machine-readable format in a common PR application. It entails clustering patterns and classifying them. Clustering created a split of the data that aids decision making, and it employs unsupervised learning. In classification, a pattern is assigned an appropriate class label based on an abstraction produced from domain knowledge or a number of training patterns. Supervised learning is employed in classification. The PR system must possess the following features, (a) unfamiliar items must be identified and classified, (b) even when patterns and objects are partially obscured, the PR system must recognize them, (c) recognize objects and shapes from different angles, (d) it should be able to recognize a pattern precisely and rapidly, (e) recognize patterns quickly, easily, and instinctively.

Algorithms can be used to detect or mathematically observe a pattern physically. Vector feature values represent a pattern in the digital world. A feature is the outcome of one or more calculated measures that quantify some of the object's important qualities. It can be discrete or continuous. PR is the process of identifying patterns and their features with the help of advanced algorithms.

In recent times, artificial intelligence (AI) plays a significant role in our daily applications. AI is a broad discipline of information technology tasked with establishing intelligent machines that can accomplish activities that would generally need human intellect. Machine learning (ML) is an AI subset that permits systems to learn and improve independently without getting to be programmed explicitly. It employs a variety of algorithms to solve problems (e.g., Naive Bayes, support vector machines, decision trees, and so on). Deep learning (DL) is a subset of ML in AI. It uses unsupervised learning algorithms to learn from unlabeled and unstructured data. In this, artificial neural networks (ANNs) are primarily used to process the data. As a result, DL is also known as deep neural network or deep neural learning.

Applying DL and ML and extracting the meaningful features from the given image or video is known as pattern recognition (PR). Through this, the efficiency and accuracy of the PR are getting higher. In the forthcoming sections, we will discuss all these aspects in detail.

4.2 Literature Review

Pattern recognition (PR) typically covers a broad spectrum of problems, and it is tough to find an approach or unified view. Fukunaga (2013), in his book, detailed the statistical PR. He explained how to apply the PR in engineering problems such as recognizing visuals, reading characters, waveform analysis, speech recognition, and modeling the brain in biology and psychology. Bishop (2006) introduces PR in the Bayesian viewpoint in his book. He represents approximate inference algorithms that permit fast approximate answers when the exact solution is not feasible. In another work (Bishop 1995), the same author detailed neural network-based PR. He also explained various types of artificial neural network configuration used in PR and provided some real-time applications. Pavlidis (2013), in his work, demonstrated structural PR and the feasibility of adopting the same in the real world. He discussed mathematical techniques for curve fitting, analytical description of region boundaries, and fundamentals of picture representation. Kumar et al. (2005) discussed correlation PR, which is a subset of statistical PR. He explained about attaining correlation PR by creating or selecting a reference signal and then determining the degree to which the object is under examination resembles the reference signal.

Shi (2018) discussed in detail fractal feature-based signal pattern recognition using machine learning (ML). He applied fractal theory for feature extraction and PR because of its extraordinary ability to express complicated information in a simple manner. Fatima et al. (2017) explained ML techniques for PR in disease diagnostic in healthcare applications. It is primarily used in telemedicine applications for diagnosing diseases. Bhamare et al. (2018) provided a comprehensive review of various ML algorithms used for PR in multiple domains. Fragassa et al. (2019) discussed ML-based PR used to predict the tensile behavior of cast alloys with experimental data. This approach is mainly used in many alloy industries. Amin et al. (2017) discussed the electroencephalogram (EEG) classification using ML with the PR approach used in the medical field. Chefrour (2019) explained incremental supervised learning algorithms and techniques for PR using ML. In this, incremental learning is an intriguing option and an open study subject that has become one of the machine learning and classification community's primary concerns. AlQuraishi (2021) discussed various ML techniques employed for PR in protein structure prediction. The neural network is used for predicting the protein structures into a finely resolved one from the already known structures. Kaur et al. (2021) explained the prediction of optical character recognition using ML in their work. They introduced a new system that is more useful in machine vision inspection systems.

Deep learning (DL) is the subset of ML. It is also widely used in many PR problems in recent times. In Phinyomark and Scheme (2018), Phinyomark et al. explained electromyography (EMG) PR with the help of DL and big data. This work is a more useful one to predict the health condition of muscles and the nerve cells in a human body. Wang et al. (2019) discussed sensor-based activity monitoring using DL methods. It provides good efficiency compared with conventional PR approaches. In (Bhanu and Kumar 2017; Huang et al. 2019), the authors examined the basics and applications like biometrics using PR with DL algorithms. They provide great insight into DL methods and unsupervised learning. Ijjina et al. (2017) explained human action recognition from videos using PR and motion sequence information. They used DL algorithms for feature extraction and classification of human actions in the video sequence. Lahmiri et al. (2019) discussed the nonlinear PR used to forecast cryptocurrency with deep neural networks. They provide more efficient forecasting techniques with DL compare with conventional PR. Long et al. (2021) explained scene text recognition and detection using PR with deep neural randomized networks.

Shambour (2021) proposed a multi-criteria recommender system using DL based algorithm. This work mainly focuses and is the most useful one on speech recognition and natural language processing system. Ma et al. (2021) explained the PR technique-based sensor-enabled gas detection and recognition method using DL. Ismael et al. (2021) provided the DL approach to predict the COVID-19 from the chest X-ray images. This PR method is the most useful and economically feasible one to identify the COVID-19 in this pandemic time. Umer et al. (2021) explained about facial recognition system to identify the different expressions with the trade-offs between the DL and data augmentation features. It is mainly used in psychology-based medical domains to identify the patient mindset using their facial expression.

Elsisi et al. (2021) explored DL-based Industry 4.0 and the Internet of Things (IoT) for smart buildings based on effective energy management. Industry 4.0 is the new revolution in the modern manufacturing industry, and these technologies provide a new transformation to Industry 4.0. In (Sijini et al. 2016; Fantin Irudaya Raj and Appadurai 2021; Raj et al. 2013), the authors detailed the artificial neural network-based classifiers for switched reluctance motor drives and their control techniques, mainly used in hybrid electric vehicle (HEV). Blanco et al. (2021) used DL to classify accessible industrial control systems in order to find vulnerabilities in key infrastructures. Gampala et al. (2020) discussed image processing approaches for image deblurring using DL methodologies. It is mainly used by computer vision techniques used in many industries. Chouhan et al. (2021) briefly explained a real-time gesture-based image classification scheme using PR with deep convolutional neural network (CNN). In recent years, this work has become increasingly important in computer vision-based applications. In (Raj and Balaji 2021), the authors described how deep neural networks were used in the fault analysis and fault classification of switched reluctance motors in detail.

There are so many other similar works are reported in the literature. Therefore, we can conclude that the pattern recognition problem becomes more efficient through deep learning and machine learning algorithms. So, it can be applied to various advanced domains and is more useful in a variety of industrial applications. The present work focuses on providing additional insights and more detailed knowledge about pattern recognition (PR), different training methodologies, application of deep learning (DL) and machine learning (ML) in PR problems. It is also discussed some of PR's real-time applications in various domains and its advancements with DL and ML in detail.

4.3   Pattern Recognition (PR) Problem

Pattern recognition (PR) can be defined in numerous ways. One of the important definitions (Online et al. 2021) of PR is, "It is an information-reduction process, the assignment of logical patterns or visual to classes based on the features of the patterns and their relationship". The following is the basic pattern recognition configuration. In the input of a black box termed a pattern recognition system, one unknown object is delivered as a sequence of measurements or signals. There is a collection of predefined classes at the system's output. The system's goal is to place the object into one of the classes. There are multiple objects to be recognized in a more general situation. The classification of nearby or subsequent objects may or not be dependent under this instance (Kruegel et al. 2003). We are led to different fields of pattern recognition depending on the classes and measurements, such as document recognition and analysis, identification of time signals or clinical abnormalities in medical imaging, recognition of speaker or speech, and so on.

4.3.1   Pattern Recognition (PR) Process

Usually, the PR process happens in two different stages. The first stage is an exploratory one. In this, the algorithm is looking for different patterns. The algorithm then moves on to the descriptive part, where it begins to classify the discovered patterns. In order to extract insights, these two parts are combined. First, data needs to be collected. The data can be obtained from the different sensors or the input image or video or signals from the system in real-world applications. Then, the data is cleaned up by preprocessing it and removing the noise. Then, after ML or DL algorithm analyzes the data for significant traits or elements that are common, this phase is called feature extraction. After that, the elements are clustered or categorized; this phase is known as classification. Following that, each segment is examined for insights. Finally, the acquired expertise is put into practice. Figure 4.1 shows the

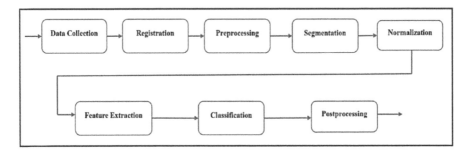

Fig. 4.1   Block diagram of the pattern recognition system

various subsections of each stage of pattern recognition. The mentioned stages may be obscure or obsolete in different types and nature of the pattern recognition system.

4.3.1.1   Collection of Data

Data collection is the first step in every PR system. Before a pattern vector can be created, a collection of measurements must be taken and transformed to numerical form with the help of some technological equipment. Scanners or video cameras are used as equipment in character recognition or image analysis; microphones are used as equipment in speech recognition, and so on. In the time or image metric domain, the input data is sampled at specific intervals, and each measurement was digitalized with a predefined number of bits. There might be a chance for additional noise and radiation signals mixed with data collected from the different fields in real time. These extra unwanted noises also create some problems to the recognition system and affect the successful PR. So, the noise needs to get removed from the system before processing further. In the upcoming stages, using filtering techniques, it will get removed. The data-gathering stage should also be built in such a way that the system can withstand fluctuations in signal measurement device operation. It needs to be operated in all types of ambient conditions and must be a robust one.

4.3.1.2   Registration

Rudimentary model fitting is done during this stage. The recognition system's internal coordinates are somehow fixed to the actual data gathered. The registration stage is designed with at least some Bayesian knowledge of the world around the system. This external data mostly answers questions like: Where does the sensible input begin, and when does it end? What method was used to gather the data? As a result, this procedure establishes the framework within which the system operates, allowing it to anticipate valid input.

4.3.1.3   Preprocessing

Real-world input data generally incorporates some noise, which necessitates some preprocessing to limit its impact. The term "noise" should be interpreted broadly: anything that prevents a PR system from performing its mission might be considered noise, regardless of how intrinsic this "noise" is in the data. Thus, before the data is given into the recognition system, preprocessing can improve some desirable aspects of the data. This stage is usually completed using a simple data filtering strategy. It is because there are so many filtering methodologies explained in the literature (Sun and Neuvo 1994; Lee et al. 1997; Haddad and Akansu 1991; Zhang et al. 2021; Javier et al. 2021; Moran et al. 2020). Many types of active and passive filters are employed in this stage. Weiner filter, Gaussian filter, and recursive filters are a few of the important filters used for filtering noise and make the input data clearer and more precise.

4.3.1.4 Segmentation

The preprocessed and registered input data must be divided into subparts that can be classified as meaningful entities. This process stage is called segmentation. It could be a distinct process or intricately entwined with subsequent or prior processes. In any instance, once the pattern recognition system has processed all of the data, the resultant segmentation of the data into its subsections can be revealed. The segmentation block can either add segment boundary information to the data flow or copy all segments into separate buffers and, one by one, pass them on to the next stage, depending on how the program was designed. Many transformation techniques are employed to segment the data in this process (Huang et al. 2020; Minaee et al. 2021; Haque and Neubert 2020; Wang et al. 2020a; Deivakani et al. 2021). Wavelet transformation (WT), discrete cosine transformation (DCT), and discrete wavelet transformation are a few of them.

4.3.1.5 Normalization

Image features that are rotation invariant are simple to define; however, the natural variance of some kinds will always elude the invariant feature extraction process. As a result, practically all pattern recognition systems require a separate normalizing phase. Loss of degrees of freedom is always an unintended consequence of normalization. It is represented in the data's intrinsic dimensionality as a reduction in dimension.

4.3.1.6 Feature Extraction

It is a dimensionality reduction technique that divides a large set of raw data into smaller groups for processing. A large number of variables in these large datasets necessitates many computing resources to process. The dimensionality of the data is reduced during the feature extraction process. It is almost necessary because of the technical limits in computation time and memory. A good feature extraction approach should enhance and preserve the characteristics of the input data that represent the different pattern classes. Simultaneously, the system should be immune to fluctuations caused by both the technical instruments and humans used during the data collecting phase. There are numerous feature extraction techniques explained in the literature (Keyvanpour et al. 2021; Yuan et al. 2020; Wang et al. 2020b; Zhou et al. 2020; Nixon and Aguado 2019; Priyadarsini et al. 2020). The selection of metrics is a problem associated with feature extraction. Individual feature variances can vary by order of magnitude, causing the classifier to fail. This problem can be solved by applying a suitable linear transform on the feature vector's components. Deep learning techniques and machine learning techniques are also employed in many recent applications for more accuracy in this stage.

4.3.1.7   Classification

It is the most crucial phase in the pattern recognition process, aside from feature extraction. All of the steps before the classification phase should be developed and refined with the classification phase in mind. There are various classifiers explained in literature for efficient classification purposes (Nakashima et al. 2007; Padol and Yadav 2016; Andaya et al. 2019; Masazhar et al. 2017; Chandra and Kaur 2021; Ngernplubpla et al. 2021). The step's operation can be summed up as a conversion of quantitative input data into qualitative output data. The classifier's outcome might be a real-valued vector giving the likelihood values for the assumptions that the pattern came from the appropriate class or a discrete picking of any one of the specified classes. Deep learning and machine learning models are used as a classifier in this phase.

The various classification algorithms are categorized into three different types: syntactic, statistical, and neural. Depending upon the type of classifier used for classification, the PR is split into three categories (Fig. 4.2). They are syntactic PR, statistical PR, and neural PR.

The statistical PR type is used to describe historical statistical data. It collects observations, processes them, and learns to generalize and apply them to new observations when learning from examples. Syntactic PR is based on simpler sub-patterns known as primitives. The pattern is described in terms of primitive connections, such

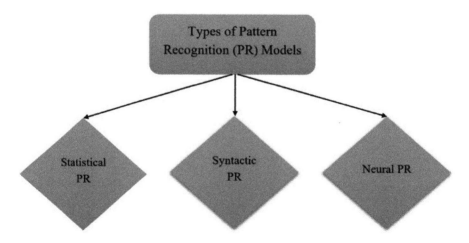

Fig. 4.2 Types of pattern recognition models

as words forming sentences and texts. Artificial neural networks are used in neural PR. They can learn complex nonlinear input-output relationships and adapt to data.

4.3.1.8 Post-processing

After the classification stage, most pattern recognition systems perform some data processing. Like the normalization processes, these post-processing subroutines introduce some prior knowledge about the surrounding world into the system. This additional knowledge can be used to improve overall classification accuracy. Inter-dependencies between individual classifications are thus resolved during this phase. It can be accomplished by running the post-processing stage alone or in conjunction with the classification and segmentation blocks. This conjunction can be attained by using loop-backs between stages.

4.3.2 Loop-Back Routes Between Stages

Figure 4.1 depicts a block diagram of an idealized PR application. In terms of accuracy of classification, such systems, in which data flows primarily from left to right, can scarcely ever be optimal. However, by allowing the subsequent blocks to interact, the system's overall performance can be considerably increased. Of course, the system gets considerably more difficult, but there is typically no other way to improve classification accuracy. Three possible backward connections paths are depicted in Fig. 4.3 with green color arrows.

The following are the reasons for these three loop-back routes: Post-processing information is provided back to the classifier. First, the post-processor alerts the classifier when it finds an impossible combination of outputs from the classifier. Either the post-processor can fix the problem independently, or it requests a new trial from the classifier. Second, the classifier revises the segmentation step. In this situation, the classifier or post-processor has identified one or more difficult-to-classify patterns. It could be a sign of misaligned segmentation, which should be identified

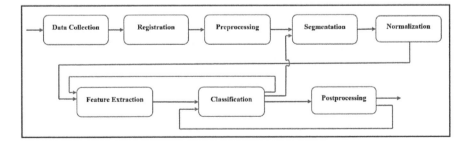

Fig. 4.3 Pattern recognition (PR) system with loop-back routes

and fixed. Finally, the feature extractor is revised based on the classifications' accuracy. This type of action is usually only achievable during the training phase, and it usually demands a classifier redesign. This method is known as error-corrective feature extraction.

4.3.3   Training Data, Testing Data, and Algorithms

Mainly, there are two types of algorithms in deep learning (DL) and machine learning (ML) used for pattern recognition. They are (a) supervised algorithms and (b) unsupervised algorithms. Mainly in ML, supervised algorithms are utilized for feature extraction or classification. On the other hand, in DL, unsupervised algorithms are also utilized for the same purpose.

4.3.3.1   Supervised Algorithms

These algorithms employ a two-stage process to identify patterns. Development of model is the first stage, and prediction of previously unseen or new things is the second stage. In this algorithm, the data is divided into two groups: testing and training. Random forest, support vector machines (SVMs), decision trees, and other machine learning algorithms can be utilized to train the model. The values in the test set have already been predicted. It is used to check the accuracy of the training set's predictions. The model's performance is measured by the number of correct predictions it makes. A classifier is a trained and tested model that uses machine learning algorithms to identify patterns.

4.3.3.2   Unsupervised Algorithms

To make a prediction, these algorithms look for patterns in the data and cluster them based on similarities in their properties. Let us assume we had a bouquet full of different fruits, such as cherries, pears, oranges, and apples. We have no idea what are fruits available in the bouquet. The data is left unmarked. Assume that someone approaches us and requests us to identify a new fruit that has just been added to the bouquet. We use a concept called as clustering in this instance. Clustering is the process of combining or grouping elements that have similar characteristics. For identifying a new item, no prior knowledge is accessible. Machine learning algorithms such as hierarchical and k-means clustering and deep learning algorithms are used in this unsupervised learning. In this, the new object is assigned to a group based on its traits or qualities in order to create a prediction. This type of training algorithms is known as an unsupervised learning algorithm.

4.3.3.3   Training and Testing Data

Learning is the process of a system being trained and becoming more versatile in order to generate accurate results. The most crucial phase of PR is learning because how well the system works with the data given is determined by the algorithms employed over it. Therefore, the whole dataset is separated into two different categories: testing the model (Testing Set) and training the model (Training Set). Figure 4.4 shows the different types of datasets and their purpose.

To make a model, you will need a training set. It is made up of the input images or other forms of inputs that are utilized for training the system. The training algorithms were used to provide vital insight into how to correlate input data to decision-making outcomes. By applying supervised algorithms on the set of data, extracting all actual information, and calculating the results, the system is trained. In most cases, 75–85% of the dataset's data is used for training. Finally, the system is tested using testing data. A dataset is used to determine whether or not a system's output is valid after it has been trained. In most cases, 15–25% of the dataset's data is used for testing. Testing data is utilized to assess the system's accuracy.

4.4   Artificial Intelligence Techniques for Pattern Recognition

Artificial intelligence (AI) is nothing but mimicking human intelligence in machines; those are programmed and designed to think like humans and perform effective operations. AI is introduced long back in history, around 1950. It comprises of expert system, fuzzy system, artificial neural network (ANN) system, etc. In recent times, the AI is reaching new heights due to the development of advanced processors and increment of memory storage device capability. It can be applied in various domains

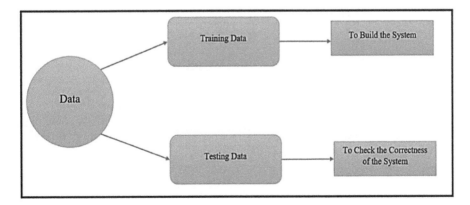

Fig. 4.4 Types of data used for learning in pattern recognition

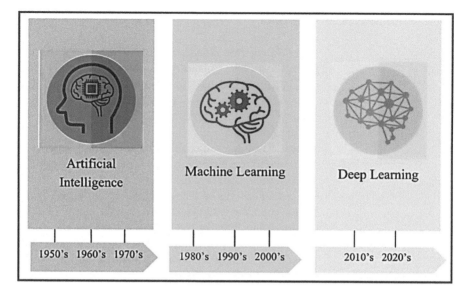

Fig. 4.5 Evolution of AI

such as robotics, computer vision, fault detection, and remote sensing. In all these applications, pattern recognition (PR) is the critical concern.

In the middle of the 1980s, machine learning (ML) is introduced. It is the subset of AI. It makes a machine capable of completing a particular task like a human with or without explicit programming. ML consists of various algorithms. Support vector machine (SVM), decision tree, linear regression, and principal component analysis are a few of them. Deep learning (DL) is the recent advent of AI. It is the subset of ML and was introduced around the year 2010. DL also makes the machine more capable of performing and executing a task with or without explicit programming. In DL, most of the algorithms learn everything on their own. Primarily ANN configurations are used in DL for classification and feature extraction. The complete evolution of DL, ML, and AI is shown in Fig. 4.5.

4.4.1 Machine Learning (ML) Techniques

As we already mentioned, ML is a subset of AI. It is the science of getting computers to perform without being explicitly programmed. This technique has given us effective Web search, self-driving cars, practical speech recognition, and a substantially enhanced knowledge of the human genome in the last decade. ML is now so common that you probably use it thousands of times a day without even realizing it. Many researchers believe it is the most effective technique to get closer to human-level AI.

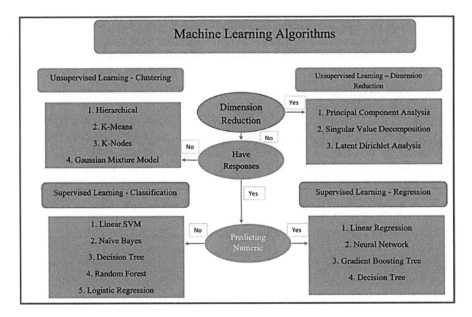

Fig. 4.6 Classification of machine learning algorithms

The ML algorithms are classified into four different categories. They are: (a) unsupervised learning—clustering, (b) unsupervised learning—dimension reduction, (c) supervised learning—classification, and (d) supervised learning—regression. Under the unsupervised learning—clustering category, there are so many algorithms listed. They are primarily used in clustering-based applications. K-means, K-nodes, and Gaussian mixture model are a few of them. The unsupervised learning—dimension reduction category consists of algorithms like principal component analysis, latent Dirichlet analysis, etc. The supervised learning category consists of various algorithms like linear SVM, Naïve Bayes, decision tree type, etc. The supervised learning—regression category consists of linear regression, decision tree, and so on. Figure 4.6 shows the detailed classification of various machine learning (ML) algorithms in detail.

In the above-mentioned ML algorithms, a few important algorithms like decision tree and SVM are used for classification and feature extraction and purpose in PR; these are explained in the following sections.

4.4.1.1 Decision Tree

PR, data mining, ML, and statistics all use decision trees as one of their predictive modeling strategies. It is created using an algorithm that looks for different methods to separate a dataset which depends upon specific criteria. Decision tree is one of the most widely used and effective supervised learning approaches. This is a supervised nonparametric learning method that can be utilized for both regression and classification. Regression trees are decision trees with a target variable that can have a range of continuous values (usually real numbers). Classification trees seem to be tree models with a discrete set of values for the target variable. The term "classification and regression tree (CART) analysis" refers to both preceding procedures.

While constructing the decision tree, we ask several types of questions at each node. The amount of information gained in relation to the question will be calculated. There are few assumptions that we made while creating the decision tree. They are (a) we consider the full training set to be the root at first, (b) the use of categorical feature values is preferred. Before the model can be constructed, the values must be discretized if they are continuous, (c) recursively, records are distributed based on attributes, and (d) we apply statistical approaches to rank attributes as the internal node or root. The general structure of the decision tree is shown in Fig. 4.7.

Decision trees are designed to replicate human decision-making abilities, making them simple to perceive. The theory underlying the decision tree is simple to understand because it has a tree-like structure. The decision tree begins at the root node. The parent node of the tree is known as the root node, while the remaining nodes are known as the child nodes. It represents the full dataset, which is then split into two or more identical groups. A tree that has been formed by splitting the tree is called a sub-tree. The process of splitting the root node/decision node into sub-nodes based on the conditions specified is known as splitting. Leaf nodes are the last output nodes, and once a leaf node has been obtained, the tree cannot be split further. The method of removing unwanted branches of the tree is called pruning.

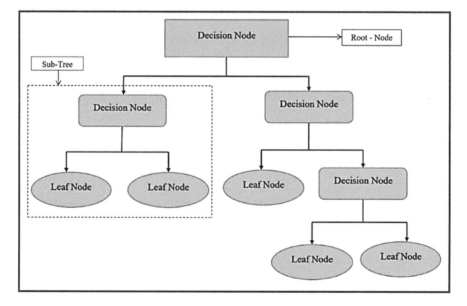

Fig. 4.7 General structure of decision tree

At each phase of the tree-building process, information collected is utilized to choose which feature to split. We want to keep our tree small since simplicity is preferable. To do so, we must select the split which generates the purest daughter nodes at each phase. The term "information" refers to a regularly used purity metric. The information value of each node in the tree reflects what more information a feature offers about the class. The procedure continues until either all of the children nodes are pure, or there is no information gain, with the first split being the one with the biggest information gain. The term "pure" denotes that all of the data in a dataset sample belongs to the same class. The term "impure" refers to a combination of different classes. Gini impurity is a statistic that measures the likelihood of a random variable being incorrectly classified for the first time if that had been arbitrarily categorized using the dataset's distribution of class labels. If our sample has a mix of classes, the chances of inaccurate categorization are considerable. If our dataset is completely clean, the chances of the wrong categorization are zero. The steps for making a decision tree are as follows:

Step 1: Obtain a list of the rows (dataset) that are considered when constructing a decision tree (recursive at each node).

Step 2: Compute the amount of data that is mixed up, the unpredictability of our dataset, the Gini impurity, and so on.

Step 3: Make a list of all the questions that must be asked at that node.

Step 4: Based on the questions asked, divide the rows into distinct as true rows and false rows.

Step 5: Calculate the information gain using the Gini impurity and the data partitioning from its preceding step.

Step 6: Update the maximum information acquired depended on each question asked.

Step 7: Depending upon a piece of new knowledge, revise the best question (higher information gain).

Step 8: Split the node to answer the best question and then restart from the beginning till we get a pure children node (leaf node).

4.4.1.2 Support Vector Machine (SVM)

Machine learning entails classifying and predicting data, and we use a variety of machine learning methods to accomplish this depending on the dataset. The support vector machine (SVM) is the highly used one in ML. It is a collection of supervised learning techniques for detecting outliers, regression, and classification. It can solve both linear and nonlinear problems and is helpful for a wide range of applications. In SVM, the algorithm generates a hyperplane or line that divides the data into categories. The extreme vectors/points that assist create the hyperplane are chosen via SVM. In n-dimensional space, there can be various lines/decision boundaries

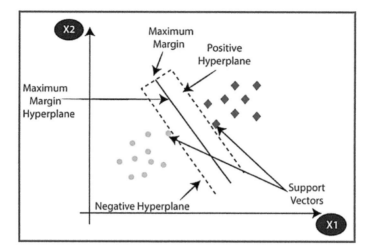

Fig. 4.8 Hyperplanes and support vector

to distinguish the classes, but we need to select the optimum decision boundary to help classify the data points. The hyperplane of SVM refers to the best boundary. Support vectors are the vectors or data points that are closest to the hyperplane and have an effect on the hyperplane's position. These vectors are called support vectors because they support the hyperplane. Consider Fig. 4.8, shown below, showing how a hyperplane or decision boundary is used to classify two groups. The features determine the hyperplane's dimensions in the dataset; for example, the hyperplane will be a straight line if there are two features (Fig. 4.8). If three features are present, the hyperplane will be a two-dimensional plane.

SVM can be categorized into two different types. They are (a) linear SVM and (b) nonlinear SVM. Only linearly separable data is used with linear SVM. For example, assume if the dataset can be classified into two categories using a straight line; then that data is known as linearly separable data. For such data classification, linear SVM is adopted. Similarly, nonlinear separable data (data cannot be categorized into two categories using a straight line) is often classified using a nonlinear SVM classifier.

First, we consider the linear SVM classifier and its working. Assume we have two datasets with two distinct features ($\times 1$ and $\times 2$) and different tags (blue and green). A classifier is required to identify whether a pair of coordinates ($\times 1$, $\times 2$) is blue or green (Fig. 4.9a). These data are linearly separable data and in two-dimensional space. Therefore, a single straight line can simply separate these two classes. Multiple lines, on the other hand, can be utilized to divide these classes (Fig. 4.9b). As a result, the SVM approach assists in determining the optimal line or decision boundary, also known as a hyperplane. The SVM method identifies the point at where the lines from both classes intersect. Support vectors are the spots where two lines intersect. The distance between the hyperplane and the vectors is called a margin. The purpose of

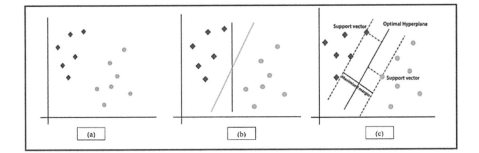

Fig. 4.9 Linear SVM

SVM is to increase this margin as much as possible. The optimal hyperplane is the one with the biggest margin (Fig. 4.9c).

We can separate data that is linearly arranged with a straight line. But we cannot separate data that is not linearly arranged using a single straight line, as shown in Fig. 4.10a. In such cases, we have to use nonlinear SVM classifier. So, in order to separate these data points, we need to add another dimension, z. We used two dimensions for linear data, x and y, so for nonlinear data, we will add a third dimension, z. It can be premeditated as $z = x^2 + y^2$. By including the third dimension, the sample

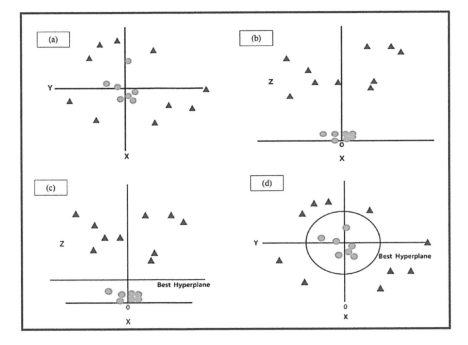

Fig. 4.10 Nonlinear SVM

space will look like this, as shown in Fig. 4.10b. As shown in Fig. 4.10c, nonlinear SVM will now divide the datasets into classes in the following manner. It appears to be a plane parallel to the x-axis since we are in three dimensions. If it is converted into 2D space with z = 1, it looks like Fig. 4.10d. We get a circumference of radius one as a result, in the case of nonlinear data.

4.4.2 Deep Learning (DL) Techniques

As we have seen already, deep learning (DL) is the subfield of ML, as shown in Fig. 4.11. It is also used in pattern recognition problems in different stages. It can act as a feature extractor or classifier. The DL enables the computer or a machine to mimic human intelligence with or without explicit training. DL is dealt with artificial neural networks (ANN), which are algorithms inspired by the function and structure of the human brain. Mostly, DL used multi-layered ANN to learn from the vast amount of data.

The ANN is typically made up of a network of interconnected units or nodes. These nodes are referred to as neurons. These artificial neurons are loosely modeled after our brain's biological neurons. The weights between neurons change as an ANN learns, as does the strength of the connection. It means that given training data and a specific task, such as number classification, we are looking for a specific set of weights that will allow the neural network to perform the classification. Each task and dataset requires a unique set of weights. We cannot predict the values of these weights ahead of time; instead, the neural network must learn them. The process of

Fig. 4.11 Artificial intelligence (AI) and its subfields

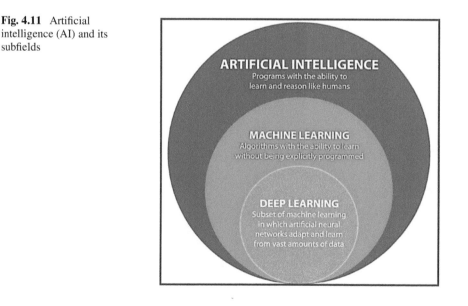

learning is also referred to as training. There are numerous ANN configurations used in DL; deep convolutional neural Network (DCNN) is among them. In the subsequent section, we will discuss DCNN in detail.

4.4.2.1   Deep Convolutional Neural Network (DCNN)

Recent improvements in convolutional neural networks (CNNs) have yielded promising results in challenging deep learning tasks. The success of a CNN, on the other hand, is contingent on finding an architecture that is appropriate for a given task. It is mainly used in computer vision-based applications, medical image processing applications, and remote sensing applications as a classifier and features extractor. It is an algorithm that takes an input image and assigns relevance (learnable biases and weights) to different elements or objects in the image, enabling it to differ-entiate between them. The degree of preprocessing required by a DCNN is much less than that required by other classification algorithms. While filters are crafted in basic approaches, DCNN can study these filters and their characteristics with minimum training. The DCNN architecture is inspired by the visual cortex structure and is analogous to the neuron connectivity pattern in the human brain. Individual neurons can only react to stimuli in the receptive field, a tiny portion of the visual field. To span the entire visual field, several comparable fields can be piled on top of one another. The simple architecture of DCNN is shown in Fig. 4.12.

The structure of DCNN, which is shown in the above Fig. 4.12, consists of convo-lutional, pooling, and fully connected layers. The convolutional layer comprises a collection of learnable kernels or filters designed to extract local characteristics from the input. A feature map is calculated for each kernel. Only a limited portion of the input, known as the receptive field, can be connected by the feature map units. A new feature map is often constructed by sliding a filter over the input and computing the dot product (which is comparable to the convolution operation) to include nonlin-earity into the model, followed by a nonlinear activation function. Each feature map

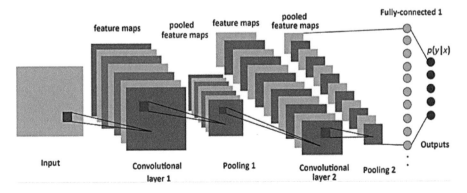

Fig. 4.12 Structure of DCNN

has the same weights (filters) for all units. The benefit of sharing weights is that it reduces the number of parameters required and allows you to detect the same feature regardless of where it appears in the inputs (Abd El Kader et al. 2021).

ReLU, tanh, and sigmoid are some of the nonlinear activation functions available. ReLU $[f(x) = \max(0, x)]$ is preferred since it speeds up training compared to the others. When we convolve an input picture with a size of $(H \times H)$ across a filter with a size of $(F \times F)$ and a stride of (S), the output size of $(W \times W)$ is provided by $\left[W = \frac{H-F}{S} + 1\right]$. The down-sampling layer or pooling layer reduces the previous feature maps' resolution. It produces very small invariance to the distortion or transformation.

Pooling can be done on a maximum or average basis. It divides the inputs into disjoint areas with a size of $(R \times R)$ to produce one output from each zone. If the pooling layer receives an input of size $(W \times W)$, the output size will be $P = W/R$. DCNNs' top layers are one or more fully connected layers, comparable to a feed-forward neural network, that aim to extract the inputs' global features. These layers' units are linked to all hidden units in the previous layer.

Typically the DCNN can be trained and optimized by using the back-propagation algorithm. Figure 4.13 shows the flowchart of back-propagation algorithm training. There are some other algorithms also suggested in the literature to train DCNN. The main drawback of DCNN is that it takes some more additional time compared with feed-forward or any other neural networks due to its greater number of hidden layers. It can adopt multi-class classification strategy like SVR along with DCNN to improve its performance. Some other hybrid strategies can also be adapted to optimize the DCNN. Figure 4.14 depicts the SVR based DCNN for diagnosis of rotary machinery (You et al. 2017).

4.5 Component of Pattern Recognition (PR) System in Real World

A PR system in the real world gets input from the sensors placed in different spots. It can work with any type of data. The data may be text, number, video, image, and so on. Figure 4.15 shows the components of the PR system in real-world applications.

After receiving the input from the sensors from the real world, the algorithms perform preprocessing. In this process, filters and segmentation algorithms are employed. First, filters remove the noise in the input data and make it clearer. Then, the segmentation algorithm segmented the input and preserved the important features in the input data. After that, the feature extraction technique is employed. It will extract the exact features from the input and provide them to the classifier. In classifier, we are using classification algorithm, cluster assignment, and regression algorithm. If it is a classifier algorithm, the class assignment will be the output.

Similarly, cluster assignment will be the output of the clustering algorithm, and the values predicted will be the output of the regression algorithm. These outputs

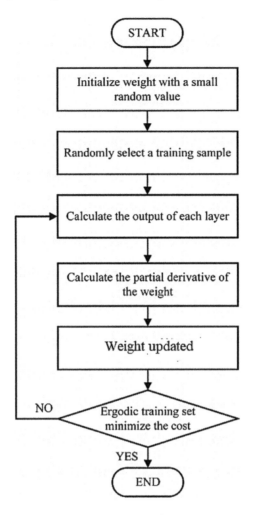

Fig. 4.13 Flowchart of back-propagation training algorithm

can be given as feedback to the preprocessing and enhancement stage or feature extraction stage, or classifier stage to increase prediction accuracy and efficiency. In this, feature extraction and classifier plays a vital role. These two can be carried out using a machine learning algorithm or a deep learning-based algorithm.

4.6 Scope and Applications of PR in Different Domains

Pattern recognition is the basis behind various domains and different applications. In this section, some of the crucial areas where PR plays an energetic role is listed.

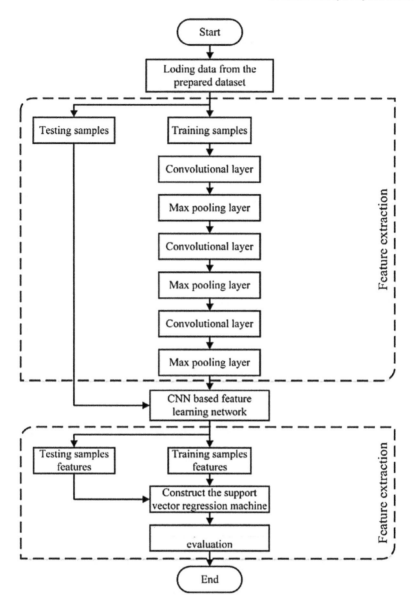

Fig. 4.14 Flowchart of SVR based CNN algorithm

1. Image Processing

Digital and analog image processing are the two types of image processing. Intelligent machine learning techniques are used in digital image processing to improve the quality of images collected from distant sources such as satellites.

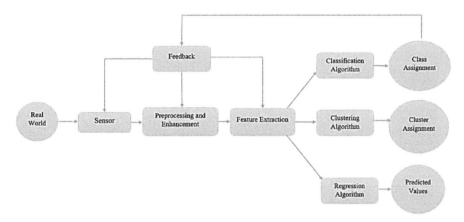

Fig. 4.15 Components of PR system in real-world applications

2. Stock Market Analysis

The stock market is challenging to forecast. Even yet, some patterns can be identified and utilized. AI is used in modern investor apps to deliver consulting services to its users.

3. Computer Vision

PR is used in computer vision techniques for extracting significant features from video or image data. In addition, it is utilized in biomedical and biological imaging, among other applications.

4. Bio-Informatics

It is a branch of research that makes predictions based on biological data using computational tools and software. Consider the case of someone discovering a new protein in the laboratory but without knowing its sequence. The unknown protein is compared to many proteins kept in a database using bio-informatics algorithms to predict a sequence based on comparable patterns.

5. Speech Recognition

Pattern recognition paradigms have yielded the most effectiveness in speech recognition. It' is utilized in various voice recognition algorithms that try to circumvent the issues that come with employing a phonetic level of depiction by treating larger units like words as patterns.

6. Recommender System

The majority of online shopping platforms uses recommender systems. These systems collect information about each consumer purchase and use machine learning algorithms to provide recommendations based on the patterns of client purchases.

7. Radar signal classification and analysis

Many applications of radar signal classifications, such as AP mine isolation and monitoring, use PR and signal processing algorithms.

8.   Medical Diagnosis

Pattern recognition algorithms that have been trained on real data can be used to diagnose cancer. These researchers have suggested an automatic breast cancer detection method with a 99.86% prediction accuracy. They extracted features from histopathology images obtained from biopsy and utilized an artificial neural network to get the results.

10.   Data Mining

It is the process of extracting valuable information from enormous amounts of data from various sources. Predictions and data analyses are based on the valuable data gathered through data mining techniques.

11.   Finger Print Identification

Fingerprint recognition is the most widely used biometric technique. Various recognition methods have been explored, even though pattern recognition approaches are the most popular to achieve fingerprint matching.

4.7   Important PR Tools Used in Recent Times

There are so many vital tools present in pattern recognition in real-time analysis. Some of the essential and famous tools are listed out here. They are,

1.   R-Studio

For code development, it uses the R programming language. Thus, it is an all-in-one development and testing environment for pattern recognition models.

2.   Microsoft Azure Machine Learning Studio

Microsoft is the one who provides it. The machine learning models are built and deployed utilizing a drag-and-drop approach. In addition, it provides a graphical user interface (GUI)-based environment for model development and usage.

3.   Google Cloud AutoML

This technology is used to create high-quality machine learning models that meet the bare minimum requirements. It builds models based on neural networks (RNNs, or recurrent neural networks) and reinforcement learning.

4.   IBM Watson Studio

IBM Watson Studio is an open-source data analysis and machine learning platform developed by IBM. It is used to build and deploy machine learning models on a desktop computer.

5.   Amazon Lex

It is an Amazon-provided open-source software/service for creating intelligent conversation agents like chatbots utilizing text and speech recognition.

4.8   Summary and Conclusion

The present work discussed pattern recognition (PR) process and its various stages in detail. In the literature review section, important works in the PR process in recent times are mentioned. We also discussed the generic PR system and its various processing stages. The loop-back routes in the generic PR system increased the efficiency and the prediction rate of the classification, and it is also getting explained. The advent of artificial intelligence (AI) with deep learning (DL) and machine learning (ML) provides new insight into the PR process. We detailed AI, ML, and DL with their various famous and important algorithms in the present study. The different types of learning algorithms and multiple components of PR systems in real world are also explained. Finally, the current work provides a list of important PR tools used in recent times. With this, we conclude that introducing new algorithms in DL and ML makes the pattern recognition process more effective. Applying these to real-world applications like computer vision, robotics, remote sensing, etc., makes the world more innovative and newly advanced.

References

Abd El Kader I, Xu G, Shuai Z, Saminu S, Javaid I, Salim Ahmad I (2021) Differential deep convolutional neural network model for brain tumor classification. Brain Sci 11(3):352
AlQuraishi M (2021) Machine learning in protein structure prediction. Curr Opin Chem Biol 65:1–8
Amin HU, Mumtaz W, Subhani AR, Saad MNM, Malik AS (2017) Classification of EEG signals based on pattern recognition approach. Front Comput Neurosci 11:103
Andaya AE, Arboleda ER, Andilab AA, Dellosa RM (2019) Meat marbling scoring using image processing with fuzzy logic based classifier. Int J Sci Technol Res 8(08):1442–1445
Bhamare D, Suryawanshi P (2018) Review on reliable pattern recognition with machine learning techniques. Fuzzy Inf Eng 10(3):362–377
Bhanu B, Kumar A (eds) (2017) Deep learning for biometrics. Springer, Switzerland
Bishop CM (1995) Neural networks for pattern recognition. Oxford University Press
Bishop CM (2006) Pattern recognition and machine learning. Springer
Blanco-Medina P, Fidalgo E, Alegre E, Vasco-Carofilis RA, Jañez-Martino F, Villar VF (2021) Detecting vulnerabilities in critical infrastructures by classifying exposed industrial control systems using deep learning. Appl Sci 11(1):367
Chandra S, Kaur M (2021) Addition of image processing techniques to our AI classifier in order to investigate the quality of utensils by composition analysis. In: 2021 5th International conference on intelligent computing and control systems (ICICCS), pp 1–8. IEEE
Chefrour A (2019). Incremental supervised learning: algorithms and applications in pattern recognition. Evol Intell 1–16

Chouhan AS, Purohit N, Annaiah H, Saravanan D, Raj EFI, David DS (2021) A Real-Time gesture based image classification system with FPGA and convolutional neural network. Int J Modern Agric 10(2):2565–2576

Deivakani M, Kumar SS, Kumar NU, Raj EFI, Ramakrishna V (2021) VLSI implementation of discrete cosine transform approximation recursive algorithm. J Phys Conf Ser 1817(1):012017. IOP Publishing

Elsisi M, Tran, M-Q, Mahmoud K, Lehtonen M, Darwish MMF (2021) Deep learning-based industry 4.0 and internet of things towards effective energy management for smart buildings. Sensors 21(4):1038

Fantin Irudaya Raj E, Appadurai M (2021) The hybrid electric vehicle (HEV)—an overview. In: Kamaraj V, Ravishankar J, Jeevananthan S (eds) Emerging solutions for e-mobility and smart grids. Springer Proceedings in Energy. Springer, Singapore.

Fatima M, Pasha M (2017) Survey of machine learning algorithms for disease diagnostic. J Intell Learn Syst Appl 9(01):1

Fragassa C, Babic M, Bergmann CP, Minak G (2019) Predicting the tensile behaviour of cast alloys by a pattern recognition analysis on experimental data. Metals 9(5):557

Fukunaga K (2013) Introduction to statistical pattern recognition. Elsevier

Gampala V, Kumar MS, Sushama C, Raj EFI (2020) Deep learning based image processing approaches for image deblurring. Mater Today Proc

Haddad RA, Akansu AN (1991) A class of fast Gaussian binomial filters for speech and image processing. IEEE Trans Signal Process 39(3):723–727

Haque IRI, Neubert J (2020) Deep learning approaches to biomedical image segmentation. Inf Med Unlocked 18:100297

Huang K, Hussain A, Wang QF, Zhang R (Eds) (2019) Deep learning: fundamentals, theory and applications (vol 2). Springer

Huang L, He M, Tan C, Jiang D, Li G, Yu H (2020) Jointly network image processing: multi-task image semantic segmentation of indoor scene based on CNN. IET Image Process

Ijjina EP, Chalavadi KM (2017) Human action recognition in RGB-D videos using motion sequence information and deep learning. Pattern Recogn 72:504–516

Ismael AM, Şengür A (2021) Deep learning approaches for COVID-19 detection based on chest X-ray images. Expert Syst Appl 164:114054

Javier MM, Galván CG, Lopez RL, Debayle J (2021) On the properties of some adaptive morphological filters for salt and pepper noise removal. Image Anal Stereology 40(1):29-à

Kaur G, Garg T (2021) Machine learning for optical character recognition system. Mach Vis Inspection Syst Mach Learn-Based Approaches 2:91–107

Keyvanpour MR, Vahidian S, Mirzakhani Z (2021) An analytical review of texture feature extraction approaches. Int J Comput Appl Technol 65(2):118–133

Kruegel C, Mutz D, Robertson W, Valeur F (2003) Bayesian event classification for intrusion detection. In 19th Annual computer security applications conference 2003, Proceedings, pp 14–23. IEEE

Kumar BV, Mahalanobis A, Juday RD (2005) Correlation pattern recognition. Cambridge University Press

Lahmiri S, Bekiros S (2019) Cryptocurrency forecasting with deep learning chaotic neural networks. Chaos, Solitons Fractals 118:35–40

Lee CS, Kuo YH, Yu PT (1997) Weighted fuzzy mean filters for image processing. Fuzzy Sets Syst 89(2):157–180

Long S, He X, Yao C (2021) Scene text detection and recognition: the deep learning era. Int J Comput Vision 129(1):161–184

Ma D, Gao J, Zhang Z, Zhao H (2021) Gas recognition method based on the deep learning model of sensor array response map. Sens Actuators B Chem 330:129349

Masazhar ANI, Kamal MM (2017) Digital image processing technique for palm oil leaf disease detection using multi-class SVM classifier. In: 2017 IEEE 4th International conference on smart instrumentation, measurement and application (ICSIMA), pp 1–6. IEEE

Minaee S, Boykov YY, Porikli F, Plaza AJ, Kehtarnavaz N, Terzopoulos D (2021) Image segmentation using deep learning: a survey. IEEE Trans Pattern Anal Mach Intell

Moran S, Marza P, McDonagh S, Parisot S, Slabaugh G (2020) DeepLPF: deep local parametric filters for image enhancement. In: Proceedings of the IEEE/CVF conference on computer vision and pattern recognition, pp 12826–12835

Nakashima T, Schaefer G, Yokota Y, Ishibuchi H (2007) A weighted fuzzy classifier and its application to image processing tasks. Fuzzy Sets Syst 158(3):284–294

Ngernplubpla J, Warunsin K, Chitsobhuk O (2021l) The performance of machine learning on low resolution image classifier. In: 2021 7th international conference on engineering, applied sciences and technology (ICEAST), pp 97–100. IEEE

Nixon M, Aguado A (2019) Feature extraction and image processing for computer vision. Academic press

Britannica Online. Encyclopedia Britannica on the Internet, 2021. Available at https://www.britan nica.com/technology/information-processing/Image-analysis

Padol PB, Yadav AA (2016) SVM classifier based grape leaf disease detection. In: 2016 Conference on advances in signal processing (CASP), pp 175–179. IEEE

Pavlidis T (2013) Structural pattern recognition (vol 1). Springer

Phinyomark A, Scheme E (2018) EMG pattern recognition in the era of big data and deep learning. Big Data Cognitive Comput 2(3):21

Priyadarsini K, Raj EFI, Begum AY, Shanmugasundaram V (2020). Comparing DevOps procedures from the context of a systems engineer. Mater Today Proc

Raj EFI, Balaji M (2021) Analysis and classification of faults in switched reluctance motors using deep learning neural networks. Arab J Sci Eng 46(2):1313–1332

Raj, EFI, Kamaraj V (2013) Neural network based control for switched reluctance motor drive. In 2013 IEEE international conference on emerging trends in computing, communication and nanotechnology (ICECCN), pp 678–682. IEEE

Shambour Q (2021). A deep learning based algorithm for multi-criteria recommender systems. Knowl-Based Syst 211:106545

Shi CT (2018) Signal pattern recognition based on fractal features and machine learning. Appl Sci 8(8):1327

Sijini AC, Fantin E, Ranjit LP (2016) Switched reluctance motor for hybrid electric vehicle. Middle-East J Sci Res 24(3):734–739

Sun T, Neuvo Y (1994) Detail-preserving median based filters in image processing. Pattern Recogn Lett 15(4):341–347

Umer S, Rout RK, Pero C, Nappi M (2021) Facial expression recognition with trade-offs between data augmentation and deep learning features. J Ambient Intell Humanized Comput 1–15.

Wang J, Chen Y, Hao S, Peng X, Hu L (2019) Deep learning for sensor-based activity recognition: a survey. Pattern Recogn Lett 119:3–11

Wang Z, Wang E, Zhu Y (2020a) Image segmentation evaluation: a survey of methods. Artif Intell Rev 53(8):5637–5674

Wang Y, Yang Y, Zhang P (2020) Gesture feature extraction and recognition based on image processing. Traitement du Signal 37(5)

You W, Shen C, Guo X, Jiang X, Shi J, Zhu Z (2017) A hybrid technique based on convolutional neural network and support vector regression for intelligent diagnosis of rotating machinery. Adv Mech Eng 9(6):1687814017704146

Yuan H, Li J, Lai LL, Tang YY (2020) Low-rank matrix regression for image feature extraction and feature selection. Inf Sci 522:214–226

Zhang L, Wang X, Dong X, Sun L, Cai W, Ning X (2021) Finger vein image enhancement based on guided tri-Gaussian filters. ASP Trans Pattern Recognit Intell Syst 1(1):17–23

Zhou W, Gao S, Zhang L, Lou X (2020) Histogram of oriented gradients feature extraction from raw Bayer pattern images. IEEE Trans Circuits Syst II Express Briefs 67(5):946–950

Chapter 5
Brain Tumor Classification Using Hybrid Artificial Neural Network with Chicken Swarm Optimization Algorithm in Digital Image Processing Application

Kalimuthu Sivanantham, I. Kalaiarasi, and Bojaraj Leena

Abstract In recent scenarios, there is a huge image processing requirement in various applications, namely pattern recognition, Image compression, multimedia computing, remote sensing, secured image data communication, biomedical imaging, and content-based image restoration. Medical image processing is the process and technique where the human body images are created for the purpose of the medical field to examine, reveal, or diagnose diseases. The medical imaging technique visualizes the internal anatomy of the human body without opening the body. The following are some of the key medical imaging techniques: X-ray, CT, PET, gamma, ultrasound, etc. The proposed research consists of various steps preprocessing used to remove noise, lung CT slices that a radiologist diagnoses are segmented using basic thresholding and morphological operations to extract the lung parenchyma. Next, the ROIs of pleural effusion are extracted, followed by the extraction of the ROIs of pneumothorax. The ROIs extracted ten shapes and texture features: convex area, equivalent diameter, mean, eccentricity, solidity, perimeter, entropy, smoothness, and standard deviation. The extracted features are applied to the ANN for training. The ANN is trained to identify the feature vectors belonging to the four class's pleural effusion, pneumothorax, normal lung, and chest CT slices affected by other diseases. When the query CT slice is applied, based on the training received, this classifies the query slice into the two classes for pneumonia or not. The classified result parameter is optimized using the chicken swarm optimization algorithm (CSO). CSO algorithm, a non-greedy local heuristic approach, is used to solve optimization issues. The optimization results exhibit an accuracy of 96.22% for pleural effusion and 97.70%.

K. Sivanantham (✉)
Crapersoft, Coimbatore, Tamilnadu, India
e-mail: sivanantham.kalimuthu@crapersoft.com

I. Kalaiarasi
Bharathiar University, Coimbatore, Tamilnadu, India

B. Leena
KGiSL Institute of Technology, Coimbatore, Tamilnadu, India

Keywords Computed tomography · Artificial neural network · Chicken swarm optimization · PCA algorithm · Feature extraction · Lung diseases · Application of medical Imaging

5.1 Introduction

The lungs are the most important organs of the respiratory system, whose main function is to draw in oxygen during inspiration and to breathe out carbon dioxide during expiration. The average inhaling rate of children is 12–20 breaths/mines. The human body requires oxygen both for cellular growth and for cellular metabolic activities. Simultaneously carbon dioxide that is a toxic waste product that has to be eliminated from the body. Malfunctioning of the lungs is referred to as lung disease. Lung diseases are one of the major health issues as they cause around four million deaths worldwide every year. In India, lung diseases are the second-highest cause of death, the major cause being coronary lung diseases (WHO 2013). India tops the world ranking for mortality due to lung diseases (Bartholmai et al. 2013). Digital and analog image processing are the two classes of image processing. Using computers, the digital images are manipulated, and hence, it is called digital image processing. Hard copies like photographs and printouts are used in analog image processing. The image processing has the following basic steps;

- Image rendering
- Image recognition
- Image description
- Image representation
- Image segmentation
- Image preprocessing
- Image acquisition
- Image classification

The image interpretation or rendering is used by the object recognized for assigning the meaning. The descriptor delivers information depending on the item that is assigned by a label in image recognition. The image description excerpts the difference between the objects of one class to another feature or quantitative information of interest features. The image representation means that the input data is transformed into a format that a computer can process. The image segmentation can divide the image (input) into their component objects or parts. Image pre-processing improves image quality and raises the likelihood of a successful process. In image acquisition, digital images are acquired. The applications of image processing are as follows:

- Remote sensing
- Morphological image processing
- Microscope image processing
- Medical image processing

- Non-photorealistic rendering
- Lane departure warning system

In image processing applications, the medical image processing application is used in our work. The ultrasound is a key medical imaging technique used to imaging soft tissues and organs. The ultrasound technique is a radiation-free, non-ionizing, and non-invasive technique. There are four major areas in medical image processing (Smistad et al. 2015). They are:

- Image creation
- Image visualization
- Image examination
- Image management

The image formation can form a digital image matrix by capturing the image steps. The image analysis contains all the processing steps that are used for the interpretation abstract of biomedical images for quantitative measurements. Image management combines all the techniques which supply image data retrieval (access), arching, transmission, communication, and effective storage. In the medical field, the accuracy of the disease diagnosis plays a vital role as it leads to further patient treatment. So the prime objective dissertation is to improve the diagnosing efficiency of the medical expert system by,

- Employing feature optimization techniques to select most significant feature subset in the medical data (Sharif et al. 2020).
- Constructing various classifier models (two-class) to train and test the clinical data (Srinivas and Rao 2019)
- Optimizing classifier parameters and fuzzy rules by using single and hybrid optimization techniques (Kumar and Mankame 2020).

In this proposed research work titled as, "Analysis and investigation of digital image processing technologies with machine learning algorithms". This work, mainly focused with various image processing schemes with machine learning algorithms to increase the imperceptibility and to enhance the robustness in the medical image processing application area. The medical data classification of data feature selection is optimization problem, which is based on the principle of picking a subset of attributes which are most significant in deciding the class label. In this approach, various image processing applications are focused, namely copyright protection, medical image classification, ownership affirmation, authentication, and high degree of sturdiness. Consequently the various medical CT image classifications to be implemented and produced better efficiency algorithms are support vector machine (Zhou et al. 2006), decision tree (Naik and Patel 2014), K-nearest neighbor algorithm (Ramdlon et al. 2019), Naive Bayes (Kaur and Oberoi 2020), and artificial neural network algorithm (Sultan et al. 2019).

In this research work, Matlab with digital image processing tools is used for simulating the results, and an efficient machine learning algorithms are used for processing of CT image data. Artificial intelligence-based classification is a long-standing field showing continuous and vigorous growth. It reduces the dimension of the data and incorporates intelligent behavior into machines and software. It is an interdisciplinary field which includes a number of sciences, professions, and specialized areas of technology. Artificial intelligence assists in the decision making process by performing data collection, treatment, processing, presentation, testing, and simulating new treatments, scenarios, and devices. During training process, the presence of instances with missing values can lead to the degradation of accuracy and performance of the classification model. By dealing these missing values suitably, the performance of the model can be improved. Case deletion is a simple and commonly used missing value handling techniques used to delete the instances with missing values. In the recent development, many meta-heuristic techniques have been inspired by the nature systems, especially computation-based algorithms named the evolutionary computing. The nature-inspired meta-heuristics techniques deal with complex computational problems and mainly used to find the best optimal solution for nonlinear problems. The ANN classified result parameter is optimized using chicken swarm optimization algorithm (CSO). Because of its non-greedy nature, this algorithm can achieve the global maxima without getting struck into local ones. Figure 5.1 explains the general block diagram for CT lung cancer prediction using machine learning approach.

The purpose of the research is to develop the complete automated computer-aided tumor detection method for valid research gap in the field of medical CT image processing and to fill the gap by addressing societal problems experienced in the medical image analysis; hence, manual interaction can be reduced. Efficient analysis of medical image is performed using the various bio-inspired techniques. This research apparently introduces several combinations of optimization and clustering techniques so as to ease the diagnosis and detection of heterogeneous tumor-affected region. The reminder of this paper is organized as follows. Section 5.2 presents CT image dataset prediction and its related work; Sect. 5.3 discussed hybrid artificial

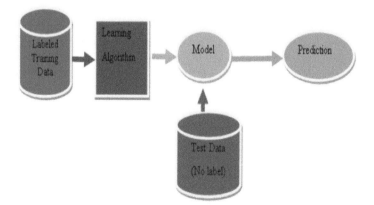

Fig. 5.1 General CT image lung cancer prediction for machine learning approach

neural network algorithm with chicken swarm optimization algorithm (HANNCSO), and Sect. 5.4 presents proposed system and existing systems' experimental results comparison. Finally, section fifth provides the concluding remarks and future scope of the work.

5.2 Literature Review

In this chapter, the various approaches of CT lungs image classification have been discussed. We have briefed some of the MR lungs image classification approaches based on nature-inspired meta-heuristics, clustering techniques, and hybrid meta-heuristic approaches. Many recent types of research are focused on secure classification due to the vastly rising development of Internet usage throughout the information world. Some of the recent advancements in image classifications are the development of hybrid algorithms, which combine the working principles of meta-heuristic with other optimization approaches and are capable to deliver different CT image classification results. It has been observed that different combinations of meta-heuristic techniques have been tried under this category, which would be of great beneficiary to clinical oncology explained in Zhou et al. (2006).

It is necessary to apply a suitable preprocessing method for the eradication of the incidence of noise conferring in the scan images. To obtain the successful results in image analysis, the quality of the image is a matter of great concern. Thus, the images for computerized diagnosis of diseases should be of high quality. Electronic preprocessing techniques should be detached for the exclusion of noise in the system. Appropriate preprocessing and shape extraction methods are obligatory for scrutinizing CT images introduced by Polsinelli et al. (2020).

Soltaninejad et al. (2017) developed the hierarchical technique for lung tumor detection using randomized tree search (RTS). This technique detects t1-weighted and t2-weighted in FLAIR MRI image. Evaluation similarity coefficient (DOI) obtained by the technique is 0.88, 88.09%, and 0.88; it can be further extended. The pattern recognition and data mining methods employment in risk prediction systems in the domain of oncology medicine was introduced by Li et al. (2006). Conventional approaches having following restrictions, high computational complexity, multiple spectral scan, and anatomical knowledge. These restrictions had been tackled here by means of classification pattern that indirectly identify difficult nonlinear affiliations among dependent and independent variables and the capability to identify each probable interactions among predictor variables defined in Gooya et al (2012). used classification and segmentation technique for gliomas called as GLISTR. The proposed technique was used to segment the gliomas present in different sequences of MR images. The high grade gliomas located in the human lungs are detected perfectly, but the segmentation process can be applied toward low-grade gliomas.

A framework had been incorporated for easy prediction of cancer disease by Kaur (2013). The framework was generated with the principal component analysis (PCA) to mine the characteristics and mathematical pattern from which to choose the related restraint. The projected work had assisted in enhancing the efficacy, precision, and process speed in feature extraction. This could apply in the applications like information retrieval, image processing, and pattern matching.

Zhuang (2016) utilized learning method dependent upon super voxel for tumor segmentation and the multi-sequence MRI images. The algorithm provides good segmentation and delineation across all grades of tumor present in MR lungs image. The automatic tumor segmentation technique produces a dice score of 0.74% against ground truth can be further extended. Rattan et al. (2019) explained a system for by using significant risk factors. The projected method composed of two most unsurpassed data mining tools, namely neural networks and genetic algorithms. The executed hybrid system employed the global optimization merits of artificial neural network algorithm to initialize neural network weights. The learning was rapid, much constant, and precise when evaluated over back propagation. The system had executed in Matlab and predicted the heart disease with precision of 89%. This literature summarized disease prediction data mining techniques, feature selection techniques, classifiers techniques, and optimization techniques.

Venkatesan and Parthiban (2017) used the combination of particle swarm optimization (PSO) and kernelized fuzzy entropy for the identification of tumor in T1-W MRI images. Lavanyadevi et al. (2017) segmentation performance of the techniques may decrease while high intensity of noise along with intensity and non-uniformity (INU) object are included into the MRI data. Radhimeenakshi (2016) integrated disease dataset classes by exploiting SVM and ANN. Investigation had been completed between two schemes based on accuracy and training time premise. The dataset exploited were the Cleveland heart database and Statlog database obtained from UCI machine learning dataset vault. Here, the data were deployed into two classes in SVM and ANN. In addition, it examined both dataset performances. Li et al. (2016) used maximum a posteriori probabilistic (MAP) technique for identifying the lungs tumor. The proposed technique can be able to segment the portions of gliomas in the MRI slices. The proposed technique produced Dice Overlap Index (DOI) for tumors of high-grade data as 0.73, 0.56, and 0.5, and they can be further increased.

Huang et al. (2012) and Deepak and Ameer (2019) had demonstrated to analyze the 3D anatomical structure of gallbladder pairs. In the developed model, the dense shape registration is implemented to the shape of the liver according to its complexity in gallbladder shape. In terms of multi-scale concavity and curvatures, the gallbladder shape is implemented in feature corresponding to the semantic shape decomposition. Hence, the set of liver-gallbladder CT data is analyzed by the developed 3D anatomical model. The experimental result shows that in the developed model the data retrieval process is fast and effective in analyzing the gallbladder pairs. The model has to be improved to incorporate it with other structures for data retrieval.

Rajathi and Radhamani (2016) objectivized to optimize the study of cancer disease prognosis by using multiple data mining methods. The authors have offered a method to enhance the projected classifier pattern by feature selection. An feature selection approaches assisted to enhance the precision of every through minimizing few low ranked attributes that aided in attaining precision of 87.8, 86.80, and 79.9% in case of SMO, Naïve Bayes, and C4.5 decision tree algorithms correspondingly.

Feitelson et al. (2015) intended a system to identify the rules effectively to predict patients risk level on the basis of a provided parameter regarding their health. The rules had been prioritized on the basis of user's prerequisite. The system performance was assessed based on precision classification, and the consequences revealed that this system had greater potential to level more precisely.

However, many methods exist for this objective which may further delay the working in fully automatic mode and to deliver much accuracy. The literature survey of various segmentation techniques for CT lungs image has been done. From the survey made, the major issues in the analysis of MR lungs image have been listed below.

a. Enormous volumes of data have to be diagnosed within less/minimum time duration.
b. The final decisions of treatment/surgery have been taken by getting the suggestion from two or more experts.
c. Identification/Segmentation of various pathologies in medical image is done manually by radiologists, leading to human errors, huge expenses, and time consumption.
d. In the field of medical image analysis, several automatic techniques have been proposed by the researchers, and even though they were reliable, they have not focused on MR images with varied tumor dimensions of disproportioned boundaries with interceded noise signals. Dealing with noise images through typified and unique segmentation algorithms is the core-objective of this research.

5.3 System Design

The blooming prominence and advancements seen in datamining in the latest generation have inspired researchers to have a comprehensive investigation. There is multiples of promising datamining research issues, on which big data classification is analyzed as a major contest to focus on. Image processing is a type of signal processing which can extract information from image or can enhance the image. The input to the image processing will be an image, and its output will also be an image (Zaitoun and Aqel 2015). The image processing has the following steps;

• Image import using a tool called image acquisition tool
• Image is manipulated and analyzed
• Output based on image analysis

Early diagnosis, good health care and an active life style can help control and reduce the risk of lung disorders, while the mortality rates can be brought down if early diagnosis of the disease is possible. This proposed work is an attempt to trace the cancer infection using adult CT image. The increase in the resolution of the CT scan has resulted in a large number of CT slices per scan. Fast and reliable image analysis techniques are the need of the hour to analyze all the CT slices of each patient. This research work initiate a hybrid method named hybrid artificial neural network with chicken swarm optimization (HANNCSO) for identifying appropriate feature subsets related to target class and given to classifier model to enhance the performance. Figure 5.2 explain the block diagram for preprocessing, segmentation, PCA feature extraction, and hybrid artificial neural network with chicken swarm optimization classifications stages.

5.3.1 Preprocessing

While capturing the digital image, the image always suffers with the noise in any kinds of imaging modalities (Elangovan and Jeyaseelan [52]), since the cause of interference of the interior elements of the image with the instruments and the environmental settings wherein the images are captured. The multiplicative or preservative noises are present in the CT image, these noises are reduce the performance of algorithm. The development of abundant cutting-edge imaging techniques has a vital role in eradicating the noisy contents surfacing in the images (Park and Schowengerdt 1983). CT image is the extensively used imaging modality specifically applied as a standard investigation method aimed at capturing soft tissue images in the human body. Recently, its significant application in the recognition of swellings and cancerous tumors has made it as a significant imaging instrument for radiologists. In examining the ultrasound images for diagnosis, it is vital that the captured image must have less quantity of noise. For instance, in case of ultra-sonogram, the speckle noise is dominant and makes it dangerous for the physician for perceiving the real gray pixels from the noise content as the infinitesimal changes of gray pixels are oblique by the speckle noise content. Table 5.1 explains the lung testing dataset used with four types of classes (normal, other slices, cavitary slices, miliary slices), scans (75, 30, 40, and 20), and slices (300, 209, 247, and 150). Totally, 150 CT scan images and 906 CT slices images are used for proposed system testing. The source for the speckle noise formation includes the distance of the transducer location from the highest pressure point. Hence, it becomes difficult to acquire the data, and also the regions having identical speckle cannot be recognized. The main purpose for the removal of speckle noise is to produce imperative results in procuring reduced false positive results and ensuring the edges to appear with better pellucidity. This helps in enhancing image quality and detecting micro-calcifications in an advanced possible measure.

The chest CT dataset is preprocessed to improve the quality of the CT slices and to make the dataset more suitable for further processing stages like segmentation and ROI extraction. Preprocessing the CT slice will enhance the radiological patterns that have been obscured by noise. The CT slices contain Gaussian and random noise. Removal of noise is done by using denoising techniques. The commonly used filters for denoising are mean filters, median filters, Laplacian filters, Weiner filters, and

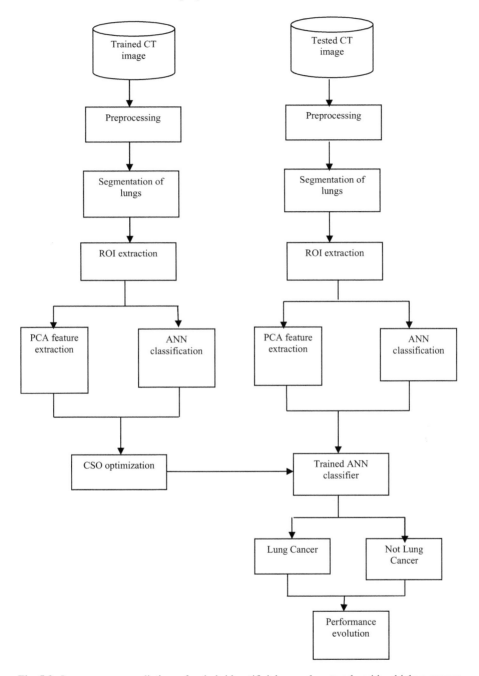

Fig. 5.2 Lung cancer predictions for hybrid artificial neural network with chicken swarm optimization (HANNCSO) using CT image dataset

Table 5.1 Dataset used for testing the proposed system

Class type	Scans	Slices
Normal	75	300
Other slices	30	209
Cavitary slices	40	247
Miliary slices	20	150
Total slices	150	906

Gaussian filters. In this research work, the Gaussian filter is used for denoising as it retains the edges in the CT slice.

The denoised and enhanced color CT slices are transformed into grayscale slices before any further processing is done. Preprocessing is an extremely important stage in the CAD system as the effectiveness of further processes such as segmentation, ROI extraction, and feature extraction is largely dependent on the enhanced quality of the dataset used. Figure 5.3 explains the input slice image and unwanted noise removed slice images differentiation. The result noise removed image attained the best quality used to improve the efficiency.

To get a better view of the abnormal masses and the anatomical structures of human lungs, contrast enhancement is applied in CT sequences of image. The most commonly used contrast agent in CT is gadlolinium. Contrast agents support in augmenting the visualization of aberrant regions from which the edema portions and malignant regions can be easily distinguished with the algorithms developed through this research. Proper ethical protocol was adopted in acquiring the images of the patient and their corresponding details. The segmentation results delivered by the algorithms developed through this thesis have been validated with the assistance of an expert radiologist and ground truth images, thus providing a clear vision regarding the need of computer-assisted diagnosis in Oncology.

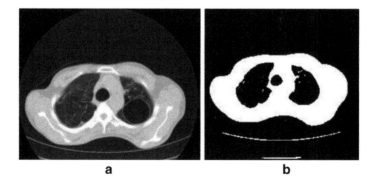

a   **b**

Fig. 5.3 **a** Input slice **b** noise removed slice

5.3.2 Segmentation

Segmentation techniques are used to partition or subdivide an image into objects. In this research work, the lung parenchyma is to be segmented from the chest CT slices. This results in the extraction of the left and the right lungs. The lungs are extracted, so that computation time need not be wasted in processing regions in the CT slice other than the lung. The chest CT slice that was transformed from RGB to grayscale in the preprocessing step is thresholded. Thresholding transforms the grayscale slice into a binary slice. The histogram of the grayscale slice is first computed, and the threshold is determined by analyzing the histogram.

Figure 5.4 explains the segmentation in five steps, background pixel removed, region growing, detect the lung slice edges, holes filled slices image generation, and finally output segmented slices image. The diseased region in the CT slice is called the region of interest (ROI) that has to be extracted. Region growing algorithm and morphological operations are versatile techniques used for ROI extraction (Dehmeshki et al. 2008). This type of radiological pattern occurs in the case of CT slices affected with military pneumonia, fibrosis, ground glass opacity, and emphysema. Hence, in the case of these diseases, region growing will not be suitable for ROI extraction. So the entire 53 lung region is divided into smaller ROI blocks of size 64×64, 32×32, or 16×16. As the block size of the ROI increases, the computations become faster as fewer ROIs have to be processed. It is inferred from literature that 32×32 and 16×16 are optimal ROI sizes. The features are extracted in the next stage from these extracted ROIs.

5.3.3 Feature Extraction

The feature extraction process extracts the shape features, texture features, and color features from the ROIs. Therefore, only the shape and texture features are extracted from the ROIs. In one of the research contributions for the classification of ILDs, the segmented lung parenchyma is divided into 32×32 sub-blocks features are extracted from the blocks using PCA algorithm (Li et al 2005). PCA algorithm following features, extraction of pleural effusion, shape and texture features like area, convex area, equivalent diameter, mean, eccentricity, solidity, standard deviation, perimeter, entropy, and smoothness are extracted from the ROIs.

5.3.4 Classification

The meta-heuristics techniques based on the population behavior take a significant role in reducing the computation time of the conventional filter as well as the machine learning-based classification. Correspondingly, the filters can harvest

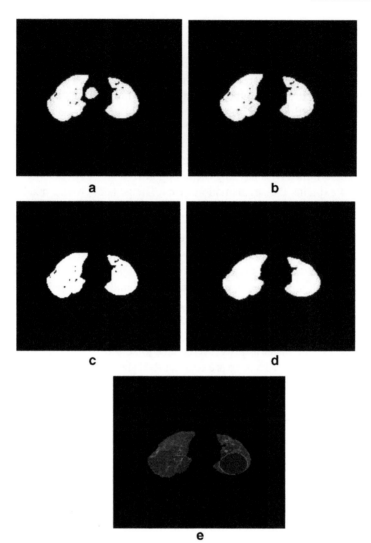

Fig. 5.4 **a** Background pixels removed, **b** Output of region growing, **c** Edge detected slice, **d** Holes filled slice, **e** Segmented slice

a smaller amount of mean square error by dint of choosing the optimal parameters with a request for a meta-heuristic algorithm. Thereby, the image quality and the image processing result can be enhanced. The classifier is trained with the feature vectors that have been extracted from the training dataset. The features are extracted from the query CT slice after preprocessing, segmentation, and ROI extraction. In this research work, the ANN classifiers (Ceylan et al. 2010) and CSO optimizers are used for classifying the query CT slices. Depending on the training received by the classifier, for the diagnosis of a specific type of disease, the features of the query slices are compared by the classifier with the corresponding features of the diseases that have already been diagnosed by the expert. In the optimization, input values provide the finest obtainable values. The process of CSO has a mathematical modeling similar to the social behavior of definite animals in the interior of their respective squad. However, CSO has been chosen due to its robustness in obtaining the best particle location globally (Kalimuthu et al. 2021).

$$v_i^l + 1 = W * v_i^l + \alpha * C_1 * \left[g_{best}^i - x_i^l \right] + \beta * C_2 * \left[p_{best}^i - x_i^l \right] \qquad (5.4.1)$$

$$x_i^{l+1} = x_i^l + v_i^{l+1} \qquad (5.4.2)$$

The assortment of v_i lies between $[v_{min}, v_{max}]$. For instance, the particle shifts when the new location is determined, and at the last iteration, the g_{best} i develops the maximal solution established.

5.4 Result and Discussion

The test image from simulation is obtained in form of JPG format with the range of 128×28 pixels applied in the experiments. It is obvious that the several multiplicative and additive noise can be reduced using the denoising work and also the filtered image conception is enhanced toward a very great extent. The proposed methodology is applied by making use of matlab2013a on Intel(R) Core (TP) i3-2410 M CPU @ 3.20 GHz and 8 GB RAM. Two different models, ANN and CSO, have achieved an average accuracy of 91.75% which were developed through diagnosis of lung cancer diseases prediction. Also compare the performance of hybrid artificial neural network with chicken swarm optimization (HANNCSO) method to other standard classifiers: support vector machine (SVM), decision tree (DT), and artificial neural network algorithm. A set of experiments is conducted on the dataset by different number of images choosen to receive the highest classification accuracy.

Table 5.2 explains the model based on proposed HANNCSO which yields the maximum false positive rate, F-score, and true positive rate for CT image dataset (81.32, 91.25, and 84.12), at most nearer artificial neural network algorithm which provides the only 54.85, 87.54, and 71.78 values; comparatively proposed approach

Table 5.2 Performance comparison lung cancer dataset

Methods	False positive	True positive	F-score
Support vector machine	31.01	78.45	54.15
Decision tree	43.65	81.87	61.78
Artificial neural network algorithm	54.85	87.54	71.78
Proposed HANNCSO system	81.32	91.25	84.12

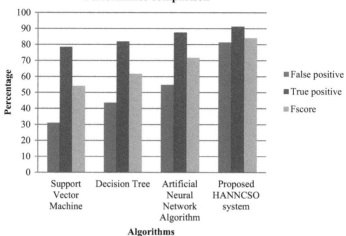

Fig. 5.5 Performance comparisons for lung cancer dataset

provides the better result in the difference terms between of 27.12, 4.25, and 12.87 values. The SVM offers the least false positive values of 31.02%, true positive values for 78.45, and F-score values 54.12 Table 5.2. In Fig. 5.5 explain the proposed hybrid artificial neural network with chicken swarm optimization (HANNCSO) and conventional algorithm performance matrix results comparison. The quality performance values exposed to the comparatively HANNCSO is better than SVM, DT, and ANN.

Table 5.3 explains the model based on proposed HANNCSO which yields the maximum efficiency, precision, and error rate for CT image dataset of (91.75, 94.65, and 6.75), at most nearer artificial neural network algorithm which provides the only 86.50 of efficiency, 90.85 of precision, and 11.24 of error rate values. Comparatively proposed approach provides the better result in the difference terms between of 5.25, 3.80, and 4.49 values.

The SVM offers the least efficiency values of 78.35%, precision for 85.6, and error rate values of 27.65 Table 5.3. While the proposed hybrid artificial neural network algorithm with ant colony optimization (HANNCSO) yields the quality matrix values for micro-array gene dataset explained in Fig. 5.6. The quality efficiency values exposed to the comparatively HANNCSO are better than SVM, DT, and ANN. The

Table 5.3 Efficiency comparison lung cancer dataset

Methods	Efficiency	Precision	Error rate
Support vector machine	78.35	85.60	27.65
Decision tree	81.25	89.70	22.68
Artificial neural network algorithm	86.50	90.85	11.24
Proposed HANNCSO system	91.75	94.65	6.75

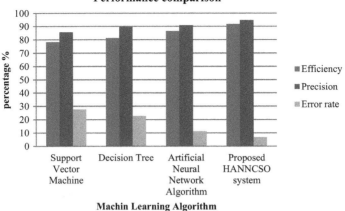

Fig. 5.6 Efficiency comparisons for ALL/AML dataset

proposed hybrid artificial neural network algorithm with ant colony optimization (HANNCSO) system unrelated and redundant structures are removed from the data, the choice of the structure will help in the enhancement of the presenting of the learning models if the decreased data goes for its classification. Proposed system achieve the following evaluation parameters accuracy 91.75, precision 94.65, error rate 06.75, false positive rate 81.32, false negative rate 91.25, and *F*-score 84.12 comparatively proposed better than existing classification method SVM and neural network.

Diagnosis The query CT slices are initially diagnosed by a radiologist to determine the type of disease. The radiologist can use the classifier results to obtain a second opinion or for confirmation of the result. The diagnosis performed by the radiologist is recorded as the CT scan report. This report is the ground truth and is used for comparison with the classifier output, wherein the radiologist diagnosed results are compared with the results of the classifier for the query CT slices. The radiologist performs reasoning with both results and arrives at the final conclusion. After comparison with the proposed standard the true positive (TP), true negative (TN), false positive (FP), and false negative (FN) values are derived.

Social Contribution Lung-based diseases; pneumonia, lung cancer, and COVID-19 are a major cause of deaths worldwide. The proposed research work will aid the radiologists in making a more accurate diagnosis. Small radiological patterns that can be missed by radiologists can still be detected by the automated computer-based CT analysis system. The proposed system enables the disease to be diagnosed at an early stage which can aid in early detection, treatment and cure of the disease which in turn can help to save human lives.

5.5 Conclusion

The prediction of diseases of the cancer is intended to help oncologist in diagnosis. This method is proposed for classifying the data on ailments in the cancer. A patient CT image classification has been performed in this work for medical dataset by using HANNCSO method, hierarchical clustering method, and max-margin classifier. This work necessitates the use of the feature selection process prior to the classification process while analyzing huge dataset. The medical history of the patients and symptoms by data mining as the data will have a major trait of selecting the method used for dataset plummeting. For this work, the method known as ANN-CSO with the micro-array gene datasets is used. A population-based system of stochastic optimization will be a healthy and an active CSO that is based on the movement of swarms. Since feeding the classifier model with entire features may cause barriers to the classification performance, a hybrid method classification is initiated to enhance the classification performance. This work is done to ascertain the imperative need for feature selection to be done prior to big data classification process and it is seen that the feature selection cannot be neglected during the classification process. The ANN training will be facilitated using the CSO for obtaining a real outcome in real-world standards. The device helps in the addressing of issues of uninterrupted augmentation. Results have proved that the HANNCSO tends to have a higher accuracy of classification by about 5.25% for the ANN, by about 9.75% for the DT and by about 13.25% for the SVM. This research summarizes with a venerable finding that utilization of ANN algorithm with CSO is the most apt among the three algorithms developed for clinical oncology, thus supporting the radiologist for better decision making in extending the life of the patients affected with different types and grades of CT cancer.

a. Heterogeneous type tumors of both primary and secondary were not segmented.
b. Significant improvement in MSE, TC, and DOI values is required.
c. Complication of analyzing huge volume of patient data due to manual interaction.

5.6 Research Scope

Similar new combinations of optimization and meta-heuristics techniques could be tried to improve the segmentation efficiency. Computational time required could be downsized considerably. The segmentation process developed through this thesis can be extended toward volume and surface rendering of anomalies present in MR slices of lungs. The application of the various techniques developed through this thesis could be used to detect and analyze the various type of pathologies found in different parts of human beings such as, kidney, liver, and lungs.

References

Bartholmai BJ, Raghunath S, Karwoski RA, Moua T, Rajagopalan S, Maldonado F, Robb RA (2013) Quantitative CT imaging of interstitial lung diseases. J Thorac Imaging 28(5)

Ceylan M, ÖZBAY Y, UÇAN ON, Yildirim E (2010) A novel method for lung segmentation on chest CT images: complex-valued artificial neural network with complex wavelet transform. Turk J Electr Eng Comput Sci 18(4):613–624

Deepak S, Ameer PM (2019) Brain tumor classification using deep CNN features via transfer learning. Comput Biol Med 111:103345

Dehmeshki J, Amin H, Valdivieso M, Ye X (2008) Segmentation of pulmonary nodules in thoracic CT scans: a region growing approach. IEEE Trans Med Imaging 27(4):467–480

Feitelson MA, Arzumanyan A, Kulathinal RJ, Blain SW, Holcombe RF, Mahajna J, Marino M, Martinez-Chantar ML, Nawroth R, Sanchez-Garcia I, Sharma D (2015) Sustained proliferation in cancer: Mechanisms and novel therapeutic targets. In: Seminars in cancer biology, vol 35. Academic Press, Dec 2015, pp S25–S54

Gooya A, Pohl KM, Bilello M, Cirillo L, Biros G, Melhem ER, Davatzikos C (2012) GLISTR: glioma image segmentation and registration. IEEE Trans Med Imaging 31(10):1941–1954

Huang W, Xiong W, Zhou J, Zhang J, Yang T, Liu J, Su Y, Lim C, Chui CK, Chang S (2012) 3D shape analysis for liver-gallbladder anatomical structure retrieval. In International MICCAI workshop on computational and clinical challenges in abdominal imaging. Springer, Berlin, Heidelberg, Oct 2012, pp 178–187

Kalimuthu S, Naït-Abdesselam F, Jaishankar B (2021) Multimedia data protection using hybridized crystal payload algorithm with chicken swarm optimization. In: Multidisciplinary approach to modern digital steganography. IGI Global, pp 235–257

Kaur AR (2013) Feature extraction and principal component analysis for lung cancer detection in CT scan images. Int J Adv Res Comput Sci Softw Eng 3(3)

Kaur G Oberoi A (2020) Novel approach for brain tumor detection based on naïve bayes classification. In: Data management, analytics and innovation. Springer, Singapore, pp 451–462

Kumar S, Mankame DP (2020) Optimization driven Deep Convolution Neural Network for brain tumor classification. Biocybernetics Biomed Eng 40(3):1190–1204

Lavanyadevi R, Machakowsalya M, Nivethitha J, Kumar AN (2017) Brain tumor classification and segmentation in MRI images using PNN. In 2017 IEEE international conference on electrical, instrumentation and communication engineering (ICEICE). IEEE, pp 1–6

Li S, Fevens T, Krzyżak A, Li S (2005) Automatic clinical image segmentation using pathological modelling, PCA and SVM. In: International workshop on machine learning and data mining in pattern recognition. Springer, Berlin, Heidelberg, July 2005, pp 314–324

Li Y, Hara S, Shimura K (2006) A machine learning approach for locating boundaries of liver tumors in CT images. In 18th International conference on pattern recognition (ICPR'06), vol 1. IEEE, pp 400–403

Li Y, Jia F, Qin J (2016) Brain tumor segmentation from multimodal magnetic resonance images via sparse representation. Artif Intell Med 73:1–13

Naik J, Patel S (2014) Tumor detection and classification using decision tree in brain MRI. Int J Comput Sci Netw Secur (ijcsns) 14(6):87

Park SK, Schowengerdt RA (1983) Image reconstruction by parametric cubic convolution. Comput Vis Graphics Image Process 23(3):258–272

Polsinelli M, Cinque L, Placidi G (2020) A light CNN for detecting COVID-19 from CT scans of the chest. Pattern Recogn Lett 140:95–100

Radhimeenakshi S (2016) Classification and prediction of heart disease risk using data mining techniques of support vector machine and artificial neural network. In: 2016 3rd International conference on computing for sustainable global development (INDIACom). IEEE, Mar 2016, pp 3107–3111

Rajathi S, Radhamani G (2016) Prediction and analysis of Rheumatic heart disease using kNN classification with ACO. In: 2016 International conference on data mining and advanced computing (SAPIENCE). IEEE, Mar 2016, pp 68–73

Ramdlon RH, Kusumaningtyas EM, Karlita T (2019) Brain tumor classification using MRI images with K-nearest neighbor method. In: 2019 International electronics symposium (IES). IEEE, Sept 2019, pp 660–667

Rattan R, Kataria T, Banerjee S, Goyal S, Gupta D, Pandita A, Bisht S, Narang K, Mishra SR (2019) Artificial intelligence in oncology, its scope and future prospects with specific reference to radiation oncology. BJRl Open 1(xxxx):20180031

Sharif M, Amin J, Raza M, Yasmin M, Satapathy SC (2020) An integrated design of particle swarm optimization (PSO) with fusion of features for detection of brain tumor. Pattern Recogn Lett 129:150–157

Smistad E, Falch TL, Bozorgi M, Elster AC, Lindseth F (2015) Medical image segmentation on GPUs–A comprehensive review. Med Image Anal 20(1):1–18

Soltaninejad M, Yang G, Lambrou T, Allinson N, Jones TL, Barrick TR, Ye X (2017) Automated brain tumour detection and segmentation using superpixel-based extremely randomized trees in FLAIR MRI. Int J Comput Assist Radiol Surg 12(2):183–203

Srinivas B, Rao GS (2019) A hybrid CNN-KNN model for MRI brain tumor classification. Int J Recent Technol Eng (IJRTE) ISSN 8(2):2277–3878

Sultan HH, Salem NM, Al-Atabany W (2019) Multi-classification of brain tumor images using deep neural network. IEEE Access 7:69215–69225

Venkatesan A, Parthiban L (2017) Medical image segmentation with fuzzy C-means and kernelized fuzzy C-means hybridized on PSO and QPSO. Int Arab J Inform Technol (IAJIT) 14(1)

Zaitoun NM, Aqel MJ (2015) Survey on image segmentation techniques. Procedia Comput Sci 65:797–806

Zhou J, Chan KL, Chong VFH, Krishnan SM (2006) Extraction of brain tumor from MR images using one-class support vector machine. In: 2005 IEEE engineering in medicine and biology 27th annual conference. IEEE, Jan 2006, pp 6411–6414

Zhuang X (2016) Multivariate mixture model for cardiac segmentation from multi-sequence MRI. In: International conference on medical image computing and computer-assisted intervention. Springer, Cham, Oct 2016, pp 581–588

Chapter 6
Detection and Classification of Breast Cancer Using CNN

R. Hariharan, M. Dhilsath Fathima, Arish Pitchai, Vibek Jyoti Roy, and Abhishek Padhi

Abstract Breast cancer has been a major type of cancer and has been responsible for countless deaths of women in the last few years. Scientists have been trying their best to come with the latest and greatest innovation in the treatment of this deadly disease. The challenge is it takes a lot of time to observe and analyze the cancer's images manually. These challenges have been overcome to a very large extent with the advancements in the research area of machine learning and deep learning. In our proposed model, ConvNet is designed to extract features from histological image of breast cancer and classifications done based upon obtained relevant feature by same network contrary to the current model where the classification is performed by using convolutional neural networks (CNN) algorithm. In our model, we have obtained 98.5% accuracy, 97.01% sensitivity and f-score 95.7%. We performed comparison with some of the existing technologies and proved that the proposed CNN has given better performance.

Keywords Breast cancer · Convolutional neural networks · BreakHis · Medical image · Classification

6.1 Introduction

The current system of treatment of breast cancer involves human monitoring and diagnosis which is time consuming and involves manual process which makes it prone to human errors. This also requires people with a very good expertise and experience. In 2018, according to WHO, breast cancer ranks second among all types of cancer in terms of claiming life's of people. So to increase the survival chances of the affected people, it is essential that the disease gets detected as early as possible. This is where this project comes into picture. Deep learning was not efficient in performance due

R. Hariharan (✉) · M. Dhilsath Fathima · V. J. Roy · A. Padhi
Department of Information Technology, Vel Tech Rangarajan Dr. Sagunthala R&D Institute of Science and Technology, Chennai, India

A. Pitchai
Quantum Scientist, Quantum Machine Learning Lab, BosonQ Psi Pvt. Ltd, Chennai, India

N. Kumar et al. (eds.), *Advance Concepts of Image Processing and Pattern Recognition*, Transactions on Computer Systems and Networks, https://doi.org/10.1007/978-981-16-9324-3_6

to lack of large dataset, but now it has become a major part of AI which is now called artificial neural network (ANN). In this paper, we have used a deep learning technique named 'convolutional neural network' for detection and classification of breast cancer in histology image (International Agency for Research on Cancer 2018). The most used deep learning model is the CNN model. In this project, we have chosen the CNN architecture for detection and classification of breast cancer using histology images from BreakHis dataset. This network architecture is used for both relevant feature extraction and classification of cancerous and non-cancerous tissue in breast. Before we apply for training, these datasets, it goes through preprocessing of images. Here we have done supervised classification that we have labeled these datasets '0' for non-cancerous and '1' for cancerous of histology images. Our proposed CNN model topology is made up of four convolutional neural network layers and one fully connected neural network layer: First layer consists of 32 channels with 3×3 filter size, second layer consists of 64 number of channels with 3×3 filter size, third layer consists of 128 channels with 3×3 filter size, fourth layer consists of 256 channels with 3×3 filter size, and fifth layer is fully connected layer. Sigmoid layer for probability distribution of classified value and, on basis of it, histology images are classified as benign breast cancer and malignant breast cancer.

6.2 Literature Survey

In field of engineering, many methods are proposed for breast cancer classification using various vision-based machine learning techniques. In this survey, we have gone through various algorithms such as K nearest neighbor (KNN), CNN, texture-based classification, Gabor wavelet transform, and every method has its own advantage and disadvantage.

Zohaib Mushtaq a et al. discussed that KNN is the most effectively used model for classification (Mushtaq et al. 2020). The performance is mainly based on the dissimilarity function being employed in the algorithm and the K value. Chi-square and L1-based selection techniques have been taken into consideration to see which one provides better performance. L1-based feature selection is the initial approach considered in this study. It gives zero weightage to features that are not significant and nonzero weightage to significant features. Models have achieved good results in terms of development by simply using KNN with an appropriate value for K and a very suitable dissimilarity function using the chi-square-based approach selection technique. A proper K value and a good approach to select the most suited distance function can fetch very good results. This method got an accuracy level of 97%. The timing for training and prediction is also low. The classification was also more efficient. However, accuracy depends upon the K value chosen.

Kui Liu et al. extracted features from image files in their article, where 2D CNN is better than MLP (Liu et al. 2018). The dense layer that comes of before the convolutional layer only works on nodes in one path, while the dense layer that comes after the convolutional layer works on all nodes in all channels. They also used a different technique, which was to use the completely connected layer as an encoder. As a result, they believe that the fully connected layer (FCLF-CNN), which utilizes the dense layer as an encoder, can outperform MLP and 1D CNN. So, two architectures have been used for building FCLF-CNN framework. They are 1D FCLF-CNN and 2D FCLF-CNN. In 2D FCLF-CNN, the dense layers that appear before convolutional layers are considered as encoder, while in the 2D FCLF-CNN, they are considered as approximator. Furthermore, for both of them, they used two separate training approaches. This is because each architecture has two loss functions: one for the layer before the convolution layer and the other for the final layer. Concurrent training applies to the whole FCLF-CNN being trained as a multi-objective model with all parameters. FCLF-CNN is designed to boost the classification efficiency of the WBCD and WDBC datasets. The advantage is before CNN use of FCL has increased the performance overall, approximator binds great relationship with target data and input data. Cons are fine-tuning undermine approximate of image data during approximation process by approximator.

Zobia Suhail et al. proposed a method for classification of the masses in terms of mammograms using texture-based approach (Suhail et al. 2018). It does not need segmentation of the mass region. Patches from filter response images were used for model generation. Texton model generation of both the classes was segregated into k number of clusters and the texton dictionary after using K-means clustering for both the benign and malignant classes. The final text on dictionary was created by combining the textons from both the classes. Fifty-three frequency histograms were produced for 53 filter responses image from the evaluation information resulting into 53 histograms of a single evaluation image. A feature vector of dimension 1060 was generated by accumulating the texton distribution of all 53 filter responses for a test ROI. The main advantage is 7×7 patch size which increases performance accuracy after many experiments, and setting $k = 10$ cluster values for this model is best. Some disadvantages are it is not suitable for large dataset, choosing right patch size for the extraction of texton from ROI is difficult, and accuracy decreases after certain clusters.

Ardalan Ghasemzadeh et al. suggested a method which stated to initially attain the feature vector appropriate to each mammography image which has been based upon Gabor wavelet transform (Ghasemzadeh et al. 2019). Several experiments are conducted for tenfold cross-validation to analyze the data complexity at each fold. The images are then cropped. The cropped images are then resized to remove noise (resized to 128×128 px).The properties and features of the Gabor wavelets are based upon human visual system and have verified to be appropriate for representing and differentiating of texture data. Gabor wavelet was used to extract features of mammography images, and the data was fed into a machine learning methods like ANN, SVM and decision trees. The accuracy level is above 0.939. The mean sensitivity is 0.951. The advantages of this proposed method are its accuracy, simplicity and robustness.

Dalal Bardou et al. proposed a method that included CNN and handcrafted features-based approach (Bardou et al. 2018; Tang et al. 2009). A bunch of local features are extracted, and these features are aggregated into an image by using a feature coding method. A spatial structure of images was matched using spatial pyramid matching. Classification of images was done using support vector machines. Optimization of the loss size was done by using stochastic gradient descent with a batch size of 32. Ordered training data was used to shuffle the dataset randomly to avoid any negative impact on the learning. Two feature-based methods were proposed. The first method includes extraction of local descriptors and encoding them with the bag of words model. The second method comprises using locality strained linear coding to encode local descriptors. Preserving the special relationship among the code vectors improves the efficiency of the classification. To improve the accuracy, they thought of using an ensemble model. A performance level of 96.15–98.33% was achieved in binary classification. The advantage includes augmentation of data for binary classification and increased accuracy. The disadvantage includes the decrease of accuracy for multiple classes due to a smaller number of training samples per class.

6.3   Methodology

In our world, people are suffering with many diseases. Some of the diseases are identified as deadly disease. In that case, breast cancer is identified as the most serious disease among women. There are many methods that have been identified to predict the cancer earlier. Even though many methods have been proposed, every method has its own pros and cons. This paper proposed a CNN-based classification. The proposed model is utilizing binary classification of histology images of breast cancer using BreakHis dataset, the method comprises convolutional neural network for extraction of feature vector from images set, and same neural network is utilized for classification purpose rather than choosing any handcrafting engineering method or techniques (Aaltonen et al. 1998).

6.3.1   Dataset Description

BreakHis (The Breast Cancer Histopathological Image Classification) is having 9109 microscopic images which is having breast images sample with positive and negative tumor identification. This data is collected from 82 patients in various magnifying factor (40X, 100X, 200X and 400X). Datasets have two different class that benign and malignant tumor. This database has created with collaboration of P&D Laboratory (Spanhol et al. 2015) (Fig. 6.4).

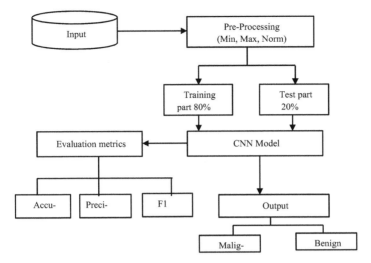

Fig. 6.1 System architecture

6.3.2   System Architecture

Figure 6.1 has shown overall work process of proposed model dataset which has been processed before splitting training and testing. 80% of data has been given for training purpose, and 20% of data has been given for testing part. After splitting, dataset is applied for designed CNN model. In CNN model 2D convolution layer, Relu layer and maxpool layer has arranged continually for feature extraction (Jafari-Marandi et al. 2018). Flatten and dense layers have been arranged, and finally output layer has been kept for final classification. Then output has been predicted as malignant or benign. Evaluation has been done based on accuracy, precision and F1-score.

6.3.3   Data Collection

The BreaKHis info contains microscopic diagnostic test pictures of efficacious as well as lethal breast tumors. Dataset has been amassed thru a scientific study from Gregorian calendar month January 2014 to Gregorian calendar month December 2014. All patents data has been taken from P&D Laboratory, Brazil which is taken to the deep study, and the verified data has been selected and created a data set. In this digital era, we are getting more digital data which helps to make more research. The institutional evaluate board permitted the study, and every affected person gave written consent. All the statistics had been anonymized. Sample's vicinity unit generated from breast tissue diagnostic check slides is stained with haematoxylin and resorcinolphthalein. The samples vicinity unit accumulated by using SOB is ready for histologic examine and labeled by means of pathologists of the P&D studies

Table 6.1 Histology image distribution

Category	Sub category	Magnification Factor			
		40×	100×	200×	400×
Benign	Adenosis	114	113	111	106
	Fibroadenoma	253	260	264	232
	Phyllodes tumor	109	121	148	115
	Tubular adenoma	149	150	140	130
Malignant	Ductal carcinoma	864	903	896	788
	Lobular carcinoma	156	170	163	137
	Mucinous carcinoma	205	222	196	169
	Papillary carcinoma	145	142	135	138

laboratory. The guidance system employed on this painting is that in the everyday paraffin technique, this is extensively hired in medical recurring. The maximum goal is to preserve the primary tissue structure and molecular composition, permitting to observe it in a completely microscope. The facts set employed on this painting is BreaKHis, which accommodates microscopic pix of diagnostic and takes a look at efficacious as well as lethal breast tumors with a complete quantity of 7909 snap shots. The information set includes eight styles of efficacious as well as lethal breast tumors. The four benign tumors sorts are glandular disorder, fibro non-malignant neoplasm, phyllodes tumors and tabular non-malignant neoplasm. The four malignant tumors sorts are ductal cancer, lobe cancer, glycoprotein cancer and papillary cancer. In the given image the area which is having data that is non-inheritable victimization is categorized as four magnification factors, those are 40X, 100X, 200X, and 400X. The amount of samples for every magnification issue is provided in Table 6.1.

6.3.4   Preprocessing

All images data are surely containing certain proportion of noisy information or redundant information which is not good for either machine learning techniques or deep learning techniques (Guo and Nandi 2006; Jiao et al. 2016). So preprocessing techniques are used to resolve those problems of images dataset in order to get better converging time and better score for proposed model. In author's proposed model, they have used min-max normalization techniques and resized from 460 × 700 to 224 × 224 as image preprocess techniques (Bae et al. 2014; Mencattini et al. 2008). This preprocessing is done in Anaconda software using Jupyter environment by Python language. In min-max normalization techniques, minimum value features get converted to zero, and maximum value features get converted to one, in range of 0 and 1. It is given by following formulae (normalized value of feature x) Eq. 6.1.

$$N = \frac{X - \min(x)}{(\max(x) - \min(x))} \tag{6.1}$$

6.3.5   CNN Model Design

Our proposed CNN model (Fig 6.2) has following topology, which consists of four CNN and one fully connected neural network, and they are given as follows, i.e., first layer consists of 32 feature maps with 3 × 3 filter size, second layer consists of 64 feature maps with 3 × 3 filter size, third layer consists of 128 feature maps with 3 × 3 filter size, fourth layer consists of 256 feature maps with 3 × 3 filter size, and fifth layer has fully connected layer and sigmoid layer (Fig. 6.3) (Hamouda et al. 2017; Jadoon et al. 2017; Shastri et al. 2018).

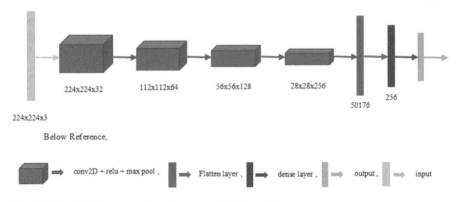

Fig. 6.2   Graphical representation of proposed CNN model

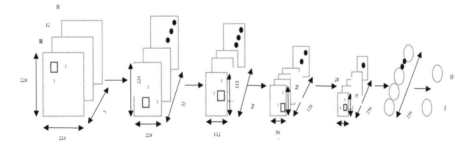

Fig. 6.3   Feature extraction of CNN model

6.3.6 Training and Testing

The BreakHis database contains training dataset of 7904 histology images, comprising 2515 benign cases and 5489 malignant cases, but in our paper, we are taking only 1500 samples in which 1000 from malignant and 500 from benign (Houssami et al. 2017; Golubnitschaja et al. 2016). These 500 samples of histology images are split into 8:2 ratio or we can say 80% for training purpose and 20% for testing purpose. These training and testing datasets feed to our proposed model the CNN architecture for further evaluation of our proposed system in order to accredit better system than existing system (Heneghan et al. 2010; Amrane et al. 2018). The processed histology images of breast cancer are passed through proposed architecture with dataset $D(k, m) = \{(an, cn); n = 1, 2, 3,.....C\}$ where n is number of samples of breast cancer histology images with corresponding classification labels (Oliveira et al. 2015; Muramatsu et al. 2016), an is nth images in dataset $D(k, m)$, and cn denotes corresponding classification labels. Generally, ConvNet neural network is represented by $X: A\ B$, where A represented images space of dataset $D(k,m)$ and B represents output probability of input images set D (Baldi et al. 2001; Dheeba et al. 2014). Then feature extraction of convolutional neural network is defined by xe and feature map of this network defined by $z = $ xe (an, ϕ| bn) where $z \in R\ w \times h \times c$, where w and h denote row and column index of sample space A, c is channel dimension of sample space of A, ϕ represents the parameter of network, and bn represents the output for each label of images samples cno (Kassani et al. 2019; Al-masni et al. 2017). Thus, operation in each layer of convolutional neural network of this architecture is calculated by $X=$ xn $- 1$, e (an, bn) $=$ Wn-1 *An $+$ Yn, where '*' represents convolutional operator, W represents weight matrix of network, and Y denotes bias matrix of network, and the final classification layer is calculated by sigmoid activation layer of network which is based on probability distribution of image vector z and defined by Eq. (6.2).

$$s = \frac{1}{(1 + e^{-z})} \tag{6.2}$$

In order to regularize or minimize our network over fitting, we have also used dropout layer in distribution of 50% before second last layer, and it has provided certain good result than before used network (Litjens et al. 2017). The cross-entropy loss function of proposed model is given by Eq. (6.3) (Fig. 6.4).

$$L = -\frac{1}{N} \sum_{n=1}^{N} c * \log(s) + (1 - c) \log(1 - s) \tag{6.3}$$

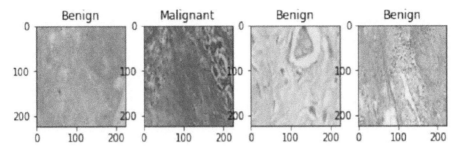

Fig. 6.4 Histology image from BreakHis dataset

6.4   Result and Discussion

The literature review discussed five advance classification techniques as follows: *K* nearest neighbor (KNN), 2D FCLF-CNN, texture-based classification, Gabor wavelet transform and handcrafted features-based classification (Arevalo et al. 2015). Major three metrics have been considered in all the discussed methods, and a value is given in Table 6.3. Resulting binary classification was divided on true positive (TP), false positive (FP), true negative (TN) and false negative (FN) regions based on confusion matrix. TP symbolizes pixels where breast cancer was identified and is really having very high possibilities. TN refers that breast cancer not being detected and also really not having cancer (Alakwaa et al. 2017; Yadav and Jethani 2016). FP is when breast cancer was detected, but really it should not be. Finally, if breast cancer was not detected, but really cancer should be present is FN. The metrics calculation is given in Table 6.2.

Table 6.2 Metric calculation

S No	Metrics	Calculation
1	Accuracy	$A = \frac{(TP + TN)}{((TP + TN + FP + FN))}$
2	Sensitivity	$Sensitivity = \frac{(TP)}{(TP + FN)}$
3	F1-score	$F1score = \frac{(2TP)}{(2TP + FP + FN)}$

Table 6.3 Metrics comparison with existing technology

Metrics	KNN	2D FCLF-CNN	Texture-based classification	Gabor wavelet transform-based classification	Handcrafted features-based classification	Proposed deep CNN
Accuracy	0.95	0.97	0.95	0.93	0.97	0.98
Sensitivity	0.96	0.96	0.97	0.95	0.95	0.96
F1-score	0.87	0.90	0.91	0.89	0.91	0.91

6.5   Conclusion

In our paper, we have performed binary classification of breast cancer using histology images from BreakHis dataset. In the proposed model, CNN architecture and network have been used for both relevant feature extraction from breast cancer histology images and classification of these images into malignant and benign based on the probability distribution of input images by sigmoid layer. This can be used for the detection of breast cancer in initial stages. We have obtained result as follows: 98.2% accuracy, sensitivity 96% and F1-score 91.7%. From this attained result, we can conclude that the proposed methodology performs better classification than existing technology. In the future, we are planning to obtain better accuracy with various deep learning models, and a hypothesis network has to be implemented for breast cancer classification and the same model can be trained and tested with a different Breast cancer dataset. Based on the metric values, best methodology will be found, and same must be optimized for better performance.

References

Aaltonen LA, Salovaara R, Kristo P, Canzian F, Hemminki A, Peltomäki P, Chadwick RB, Kääriäinen H, Eskelinen M, Järvinen H, Mecklin JP (1998) Incidence of hereditary nonpolyposis colorectal cancer and the feasibility of molecular screening for the disease. New England J Med 338(21):1481–1487

Alakwaa W, Nassef M, Badr A (2017) Lung cancer detection and classification with 3D convolutional neural network (3D-CNN). Lung Cancer 8(8):409

Al-masni, MA, Al-antari, MA, Park, JM, Gi, G, Kim, TY, Rivera, P, Valarezo, E, Han, SM, Kim TS (2017) Detection and classification of the breast abnormalities in digital mammograms via regional convolutional neural network. In 2017 39th annual international conference of the IEEE engineering in medicine and biology society (EMBC). IEEE, pp 1230–1233

Amrane M, Oukid S, Gagaoua I, Ensari T (2018) Breast cancer classification using machine learning. In: 2018 electric electronics computer science biomedical engineerings meeting (EBBT). IEEE, pp 1–4

Arevalo J, González FA, Ramos-Pollán R, Oliveira JL and Lopez MAG (2015) August Convolutional neural networks for mammography mass lesion classification. In 2015 37th annual international conference of the IEEE engineering in medicine and biology society (EMBC). IEEE, pp 797–800

Bae MS, Moon WK, Chang JM, Koo HR, Kim WH, Cho N, Yi A, La Yun B, Lee SH, Kim MY, Ryu EB (2014) Breast cancer detected with screening US: reasons for nondetection at mammography. Radiology 270(2):369–377

Baldi P, Brunak S, Bach F (2001) Bioinformatics: the machine learning approach. MIT press

Bardou D, Zhang K, Ahmad SM (2018) Classification of breast cancer based on histology images using convolutional neural networks. IEEE Access 6:24680–24693

de Oliveira FSS, de Carvalho Filho AO, Silva AC, de Paiva AC, Gattass M (2015) Classification of breast regions as mass and non-mass based on digital mammograms using taxonomic indexes and SVM. Comput Bio Medi 57:42–53

Dheeba J, Singh NA, Selvi ST (2014) Computer-aided detection of breast cancer on mammograms: a swarm intelligence optimized wavelet neural network approach. J Biomed Inf 49:45–52

Ghasemzadeh A, Azad SS, Esmaeili E (2019) Breast cancer detection based on Gabor-wavelet transform and machine learning methods. Int J Mach Learn Cyber 10(7):1603–1612

Golubnitschaja O, Debald M, Yeghiazaryan K, Kuhn W, Pešta M, Costigliola V, Grech G (2016) Breast cancer epidemic in the early twenty-first century: evaluation of risk factors, cumulative questionnaires and recommendations for preventive measures. Tumor Bio 37(10):12941–12957

Guo H, Nandi AK (2006) Breast cancer diagnosis using genetic programming generated feature. Pattern Recognit 39(5):980–987

Hamouda SKM, El-Ezz RHB, Wahed ME (2017) Enhancement accuracy of breast tumor diagnosis in digital mammograms. J Biomed Sci 6(4):1–8

Heneghan HM, Miller N, Kerin MJ (2010) MiRNAs as biomarkers and therapeutic targets in cancer. Curr Opin Pharmacol 10(5):543–550

Houssami N, Lee CI, Buist DS, Tao D (2017) Artificial intelligence for breast cancer screening: opportunity or hype? Breast 36:31–33

International Agency for Research on Cancer (2018) Global cancer observatory

Jadoon MM, Zhang Q, Ul Haq I, Jadoon A, Basit A, Butt S (2017) Classification of mammograms for breast cancer detection based on curvelet transform and multi-layer perceptron. Biomed Res (0970-938X), 28(10)

Jafari-Marandi R, Davarzani S, Gharibdousti MS, Smith BK (2018) An optimum ANN-based breast cancer diagnosis: Bridging gaps between ANN learning and decision-making goals. Appl Soft Comput 72:108–120

Jiao Z, Gao X, Wang Y, Li J (2016) A deep feature based framework for breast masses classification. Neurocomputing 197:221–231

Kassani SH, Kassani PH, Wesolowski MJ, Schneider KA, Deters R (2019) Breast cancer diagnosis with transfer learning and global pooling. In: 2019 international conference on information and communication technology convergence (ICTC). IEEE, pp 519–524

Litjens G, Kooi T, Bejnordi BE, Setio AAA, Ciompi F, Ghafoorian M, Van Der Laak JA, Van Ginneken B, Sánchez CI (2017) A survey on deep learning in medical image analysis. Med Image Anal 42:60–88

Liu K, Kang G, Zhang N, Hou B (2018) Breast cancer classification based on fully-connected layer first convolutional neural networks. IEEE Access 6:23722–23732

Mencattini A, Salmeri M, Lojacono R, Frigerio M, Caselli F (2008) Mammographic images enhancement and denoising for breast cancer detection using dyadic wavelet processing. IEEE Trans Instrum Measur 57(7):1422–1430

Muramatsu C, Hara T, Endo T, Fujita H (2016) Breast mass classification on mammograms using radial local ternary patterns. Comput Bio Med 72:43–53

Mushtaq Z, Yaqub A, Sani S, Khalid A (2020) Effective K-nearest neighbor classifications for Wisconsin breast cancer data sets. J Chin Instit Eng 43(1):80–92

Shastri AA, Tamrakar D, Ahuja K (2018) Density-wise two stage mammogram classification using texture exploiting descriptors. Exp Syst Appl 99:71–82

Spanhol FA, Oliveira LS, Petitjean C, Heutte L (2015) A dataset for breast cancer histopathological image classification. IEEE Trans Biomed Eng 63(7):1455–1462

Suhail Z, Hamidinekoo A, Zwiggelaar R (2018) Mammographic mass classification using filter response patches. IET Comput Vis 12(8):1060–1066

Tang J, Rangayyan RM, Xu J, El Naqa I, Yang Y (2009) Computer-aided detection and diagnosis of breast cancer with mammography: recent advances. IEEE Trans Inf Technol Biomed 13(2):236–251

Yadav P, Jethani V (2016) Breast thermograms analysis for cancer detection using feature extraction and data mining technique. In: Proceedings of the international conference on advances in information communication technology & computing, pp 1–5

Chapter 7
De-Noising of Poisson Noise Corrupted CT Images by Using Modified Anisotropic Diffusion-Based PDE Filter

Nikhil Singh and R. B. Yadav

Abstract In Poisson noise, the number of collected photons is so small due to the low light environment that degrades the image by reducing image resolution and contrast. Therefore, Poisson noise evaluation is necessary for the restoration of digital images. In this chapter, a filter based on the partial differential equation is introduced to restore noisy images. The proposed PDE-based filter describes Poisson noise-adapted anisotropic diffusion-based method in the L-2 framework. Regularization function and data fidelity are two components of this filter. A regularization parameter lambda is inserted throughout the filtering process to ensure proper stability between the data fidelity and the regularization function. Evaluation of performance parameters for all the described techniques is done with the help of MSE and PSNR.

Keywords Anisotropic diffusion · Computed tomography · Image reconstruction · PDE filters

7.1 Introduction

The photons released by an object are measured to determine image quality in a variety of image applications, including astronomical imaging, medical imaging and so on. For such cases, Poisson distribution models the image noise. Maximum likelihood estimation (MLE) (Srivastava and Srivastava 2013) for Poisson is introduced for retrieving the noise corrupted images (Jain 1989). When the underlying Poisson maximum likelihood model equation is ill-posed, regularization may be required.

Suppose the observed unclear and noisy picture array is represented by i. In that case, the estimation of the $Z \times Z$ true object array n is obtained by solving a linear system (Takezawa 2005; Papoulis and Pillai 2002) approximately in the form of

N. Singh
Department of Electronics and Communication Engineering, Delhi Technological University (Formerly Delhi College of Engineering), Delhi, India
e-mail: nikhil_2k19phdec04@dtu.ac.in

R. B. Yadav (✉)
Department of Electronics and Communication, G. B. Pant Institute of Engineering and Technology, Pauri Garhwal, India

N. Kumar et al. (eds.), *Advance Concepts of Image Processing and Pattern Recognition*, Transactions on Computer Systems and Networks, https://doi.org/10.1007/978-981-16-9324-3_7

$$g = \int Pn + r \qquad (7.1)$$

The common form of image restoration method reads

$$g = F(Pn + r) \qquad (7.2)$$

where g represents $Z^2 \times 1$ the size column stacked vector; $F(.)$ is the noise-making process; the unknown/known point spread function (PSF) is denoted by P; r denotes the image's positive background intensity, and Pn is positive for all $n \geq 0$.

The components of a corrupted image x pertain to the amount of noisy photons that follow a Poisson distribution in imaging applications such as astronomical imaging and medical imaging. The mathematical model which shows the inaccuracy in such system reads

$$x = \text{Poisson}(Pn + r) = \text{Poisson}(i) \qquad (7.3)$$

where $\text{Poisson}(Pn + r)$ is a Poisson random vector with Poisson parameter vector I that is independent and uniformly distributed. The probability density function (PDF) for Poisson noise (Nocedal and Wright 1999; Lane 1996) corrupted data I_o reads

$$p(n/x) = \prod_{l=1}^{n} \frac{([Pn]_l + r_l)^{(x)_l} . e^{-([Pn]_l + r_l)}}{x!} \qquad (7.4)$$

Consider the following Poisson PDF if we substitute $I_o = x$ and $I \approx Pn + r$, with, I_o being the detected image and I being the reconstructed image, we get the following Poisson PDF:

$$p(I/I_0) = \frac{e^{-I} . I^{I_o}}{I_o!} \qquad (7.5)$$

By maximizing $p(I/I_o)$ in reference to I and focusing the constraint $I \geq 0$, the maximum probability estimate of I is derived. Alternatively, we cannot get the MLE of Poisson noise by lowering the negative log-likelihood PDF (Perona and Malik 1990; Lane 1996; Llacer and Núñez 1991; Shepp and Vardi 1982) of I which is given by

$$I_{\text{ML}} = \underset{I \geq 0}{\arg \min} \{\ln p(I/I_o)\} = \underset{I \geq 0}{\arg \min} \{I - I_o \ln I\} \qquad (7.6)$$

For Poisson noise, maximum likelihood estimate (MLE) is

$$I_{\text{ML}} = \frac{(I_o - I)}{I} \qquad (7.7)$$

Equation (7.7) gives the solution for Eq. (7.6) which can be degraded by noise if PSF of P is ill-conditioned.

To get the answer or best estimate of I, the resulting optimization issue can be phrased in a Bayesian context using the maximum a posterior (MAP) method (Rudin et al. 1992) or by expressing the issue as a minimization problem in a variational framework (Lane 1996).

7.2   General Frame for CT Image Restoration

7.2.1   MAP Methodology

To resolve the issue of image reconstruction, the proposed filter utilizes the properties of the maximum a posterior (MAP) method. In Bayesian contexts, a probability density parameter $D(I)$ for I is provided, and the posterior density in relation to I is maintained as follows (Papoulis and Pillai 2002):

$$D(I/I_o) = \frac{\mathrm{d}(I_o/I)\mathrm{d}(I)}{\mathrm{d}(I_o)} \tag{7.8}$$

Probability for log-posterior maximization occurs to process the initial noise image I_o for the reconstruction of the filtered image I.

$$\log(\mathrm{d}(I/I_o))\eta \log(\mathrm{d}(I_o/I)) + \log(\mathrm{d}(I)) \tag{7.9}$$

where the proposed noisy model $\mathrm{d}(I/I_o)$ shows the likelihood term and $\mathrm{d}(I)$ shows the prior term.

When the filtering problem is framed as maximizing a posterior, the likelihood term of Poisson noise can be integrated as an attachment term that can be incorporated to an already existing model of an image. Gibb's image prior model is employed in this manuscript.

The Gibbs prior image model is incorporated in this manuscript. Expressed in terms of image gradient norms, the Gibbs prior model utilizes the energy functional and is coupled to an anisotropic diffusion-based partial differential equation (You and Kaveh 2000) (Srivastava et al. 2010b) for the removal of additive noise. The previous Gibbs model is as follows (Voci et al. 2004; Gilboa et al. 2004):

$$\mathrm{d}(I) = \exp(-\lambda Z(I)) \tag{7.10}$$

where energy functional (Bardsley and Goldes 2009) is described as follows:

$$Z(I) = \arg\min(\lambda \int_{\Omega} \varphi(\|\nabla I\|)\mathrm{d}\Omega) \tag{7.11}$$

If $\varphi(\|\nabla I\|) = \|\nabla I\|^2$ is image gradient norm, then the prior function given by the corresponding energy function reads

$$Z(I) = \arg\min(\lambda \int_\Omega \|\nabla I\|^2 d\Omega) \tag{7.12}$$

7.2.2 Minimization Framework

Maximizing the $d(I/I_o)$ in Eq. (7.8) is the same as minimizing the following:

$$T(I) = T_o(I/I_o) - \ln d(I) \tag{7.13}$$

where $T_o(I/I_o)$ is the maximum likelihood estimate of I and signifies as the likelihood parameter, and the second parameter signifies to the second function. $d(I)$ is the previous term from which the unknown I is assumed to rise, and $-\ln d(I)$ is the regularization term from typical inverse issues.

The following is the definition of regularization function:

$$-\ln p(I) = \frac{\lambda}{2}\langle CI, I \rangle \tag{7.14}$$

where $\langle CI, I \rangle$ represents the inner Euclidean product, C is the non-negative symmetric regularization function, and λ maintains equilibrium among fidelity and data regularization.

As a result, the casting of the image restoration problem is demonstrated by the minimization problem (Bardsley and Laobeul 2008):

$$I_\lambda = \arg\min_{I \geq 0}\left\{ T_\lambda(I) = T_o(I/I_o) + \frac{\lambda}{2}\langle CI, I \rangle \right\} \tag{7.15}$$

where $T_o(I/I_o)$ shows the data fidelity, and for Poisson noise, it can be denoted as $\|I - I_o \ln I\|$; the second function is the regularization function with the choices of C; the regularization function determines the amount of smoothing, and I_λ denotes the image that has been de-noised by minimizing Eq. (7.15).

Whittaker (Whittaker 1922) introduced the standard minimization problem utilizing least-square minimization, and it has been advanced by Wahba (Wahba 1990). This section modifies the traditional approach for Poisson noise in a variational framework.

7.2.3 Methods and Models

Equation (7.15) shows that the minimization problem is of variational framework and can be represented in the form of

$$I_\lambda = \arg\min_{I \geq 0} E(I) = \int_\Omega \left(T_o(I/I_o) + \frac{\lambda}{2} \langle CI, I \rangle \right) \mathrm{d}\Omega \qquad (7.16)$$

where Ω is image space.

The above equation, i.e., (7.16), can further be written as follows:

$$I_\lambda = \arg\min_{I \geq 0} E(I) = \int_\Omega (T_o(I/I_o) - \ln d(I)) \mathrm{d}\Omega \qquad (7.17)$$

Equation (7.10), i.e., Gibbs prior equation is used to reduce Eq. (7.17) to:

$$I_\lambda = \arg\min_{I \geq 0} E(I) = \int_\Omega (T_o(I/I_o) + \lambda E(I)) \mathrm{d}\Omega \qquad (7.18)$$

where the first term computes the fidelity of data and the other term is the regularization parameter.

Further for Poisson PDF, Eq. (7.18) reads

$$T_o(I/I_o) = I - I_o \ln I \qquad (7.19)$$

In a variational function, the regularization function $E(I)$ is a function of gradient norm and is denoted as follows:

$$E(I) = \varphi(|\nabla I|) \qquad (7.20)$$

7.2.4 Anisotropic Diffusion (Yu and Acton 2002)-Based Method

$CI = \|\nabla I\|$: total variation (TV) (Rudin et al. 1992).

The subsequent term, i.e., regularization parameter in Eq. (7.15), reduces to total variation regularization (Rudin et al. 1992), which is implemented in the $L1$ framework and penalized by $L1$-norm reconstructions (Srivastava et al. 2009).

Total variation penalized Poisson MLE is achieved by reducing the following:

$$I_\lambda = \arg\min_{I \geq 0}\{T_\lambda(I) = (I - I_o \ln) + \lambda |\nabla I|\} \qquad (7.21)$$

For Poisson noise-adapted AD-based filter, the regularization term $E(I) = \varphi(\|\nabla I\|)$ is defined as follows:

$$\varphi(\|\nabla I\|) = \|\nabla I\|^2, \text{(L2 norm)}, \tag{7.22}$$

As a result, the proposed TV-centered scheme reads as follows:

$$\arg\min_{I \geq 0} E(I) = \int_{\Omega} \left((I - I_o \ln I) + \lambda \|\nabla I\|^2\right) d\Omega \tag{7.23}$$

On the set of $I \in BV(\Omega)$, the function $E(I)$ is defined such that $\log I \in L^1(\Omega)$, and it is necessary to be positive in all conditions for I.

In Eq. 7.23, we get the following results using the Euler–Lagrange minimization method:

$$0 = \nabla.(c\|\nabla I\|\nabla I) + \frac{(I_o - I)}{\lambda I}, \text{ with } \frac{\partial I}{\partial \vec{n}} = 0 \text{ on } d\Omega \tag{7.24}$$

Equation 7.24 is solved using the gradient descent, and the attained result is as follows:

$$\frac{\partial I}{\partial t} = \nabla.(c\|\nabla I\|\nabla I) + \frac{1}{\lambda} \cdot \frac{(I_o - I)}{I}, \text{ with } \frac{\partial I}{\partial \vec{n}} = 0 \text{ on } d\Omega \tag{7.25}$$

where $c\|\nabla I\|$ denotes the coefficient of diffusion and is described as (You and Kaveh 2000; Srivastava 2011; Srivastava et al. 2010a)

$$c\|\nabla I\| = \frac{1}{1 + \left(\frac{\nabla I}{f}\right)^2} \tag{7.26}$$

As a result, the proposed scheme of Poisson noise-adapted anisotropic diffusion filter is given as follows:

$$\frac{\partial I}{\partial t} = \nabla.(c\|\nabla I\|\nabla I) + \frac{1}{\lambda} \cdot \frac{(I_o - I)}{I}, \text{ with } \frac{\partial I}{\partial \vec{n}} = 0 \text{ on } d\Omega \tag{7.27a}$$

with initial condition

$$I(t = 0) = I_o \tag{7.27b}$$

The first term in Eq. (7.27a) is responsible for picture regularization and smoothing by lowering pixel variance. In contrast, the second term signifies an attachment term for data (Srivastava et al. 2012), measuring the difference between the observed image and its restored value achieved through the filtering process.

Equation (7.26) gives the diffusion coefficient, which is employed in Eq. (7.27a), and the adaptive value of f is obtained utilizing robust statistic techniques (Voci et al. 2004) to impulsively approximate the robust value σ_e of an image I. The value of the f is set to σ_s, which is the image's gradient's minimum absolute difference. The modified value of f can be calculated as under:

$$f = \sigma_s = \Lambda \times \text{median}_I[\|\nabla I - \text{median}_I(\|\nabla I\|)\|] \qquad (7.28)$$

where $\Lambda = 1.4826$.

7.2.5 Digitization of the Proposed Model

For digital implementation, the Eqs. (7.27a) and (7.27b) can be made digital using finite difference technique (Llacer and Núñez 1991). Equations (7.27a) and (7.27b) demonstrate the digital version for suggested anisotropic diffusion-based model (Scherzer 2010) (Gilboa et al. 2004).

$$R^{n+1} = R^n + \Delta t.\left[\frac{1}{\lambda}.\frac{(R_o - R^n)}{R^n} + \nabla.(c\|\nabla R^n\|\nabla R^n)\right] \qquad (7.29a)$$

$$R(t = 0) = R_o \qquad (7.29b)$$

The von Neumann analysis (Llacer and Núñez 1991) shows that we require $\frac{\Delta y}{(\Delta s)^2} \langle \frac{1}{4}$ the numerical scheme, given by Eq. (7.29a) and (7.29b), to be stable. If the grid size is set to $\Delta s = 1$, then $\Delta y < 1/4$. Therefore, for the stability of Eqs. (7.29a), (7.29b) and similarly for others, the value of Δt is set to 0.25. (Srivastava et al. 2011).

7.3 Results and Discussions

In the given division, quantitative and qualitative analysis of MSE and PSNR (Shepp and Vardi 1982) for all methods has been compared for the CT abdomen (https://data.idoimaging.com/dicom/1020_abdomen_ct/1020_abdomen_ct_510_jpg.zip) image of size 512×512. It has been witnessed that up to 100 iterations, the proposed model produces satisfying results. The number of iteration for each technique was fixed to 100 as the value of PSNR starts decreasing above 100 iterations. For the condition of stability, 0.25 is the value set for Δt. The suggested filter takes a picture distorted by Poisson noise and returns the reconstructed image as input.

Figure 7.1 shows the visual outcome analysis of implemented restoration methods. Figure 7.2 compares the performance of different restoration approaches in terms

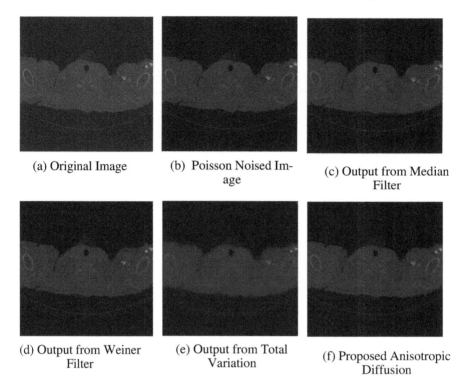

(a) Original Image   (b) Poisson Noised Image   (c) Output from Median Filter

(d) Output from Weiner Filter   (e) Output from Total Variation   (f) Proposed Anisotropic Diffusion

Fig. 7.1 Visual result comparison of different restoration methods

of MSE and PSNR, revealing that the anisotropic diffusion method outperforms the others.

7.4   Conclusion

A Poisson noise-adapted PDE filter has been proposed in this manuscript to restore and enhance CT images. The likelihood term, regularization parameter and regularization function were all observed under this scheme. The Poisson PDF maximum likelihood computation has been simplified mathematically. MATLAB was used to implement all of the methods. The performance of different image restoration approaches is measured with the help of PSNR and MSE. It may be deduced from the obtained results that the Poisson noise-adapted anisotropic diffusion method outperforms other image restoration methods. This approach is capable of restoring good image quality while maintaining strong edge preservations.

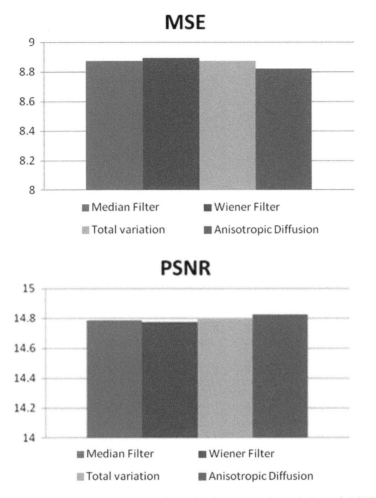

Fig. 7.2 a MSE-based performance comparison of various restoration techniques. **b** PSNR-based performance comparison of various restoration techniques

References

Bardsley JM, Goldes J (2009) Regularization parameter selection methods for ill-posed Poisson maximum likelihood estimation. Inverse Prob 25(9):095005

Bardsley JM, Laobeul ND (2008) Tikhonov regularized Poisson likelihood estimation: theoretical justification and a computational method. Inverse Prob Sci Eng 16(2):199–215

Gilboa G, Sochen N, Zeevi YY (2004) Image enhancement and denoising by complex diffusion processes. IEEE Trans Pattern Anal Mach Intell 26(8):1020–1036

https://data.idoimaging.com/dicom/1020_abdomen_ct/1020_abdomen_ct_510_jpg.zip

Jain AK (1989) Fundamentals of digital image processing

Lane RG (1996) Methods for maximum-likelihood deconvolution. JOSA A 13(10):1992–1998

Llacer J, Núñez J (1991) Iterative maximum likelihood estimator and Bayesian algorithms for image reconstruction in astronomy. In: The restoration of HST images and spectra, p 62

Nocedal J, Wright SJ (1999) Numerical optimization. Spring er-Verlag, Berlin, Heidelberg, New York

Papoulis A, Pillai SU (2002) Probability, random variables, and stochastic processes. Tata McGraw-Hill Education

Perona P, Malik J (1990) Scale-space and edge detection using anisotropic diffusion. IEEE Trans Pattern Anal Mach Intell 12(7):629–639

Rudin LI, Osher S, Fatemi E (1992) Nonlinear total variation based noise removal algorithms. Physica D 60(1–4):259–268

Scherzer O (Ed) (2010) Handbook of mathematical methods in imaging. Springer Science & Business Media

Shepp LA, Vardi Y (1982) Maximum likelihood reconstruction for emission tomography. IEEE Trans Med Imaging 1(2):113–122

Srivastava R, Srivastava S (2013) Restoration of Poisson noise corrupted digital images with nonlinear PDE based filters along with the choice of regularization parameter estimation. Pattern Recogn Lett 34(10):1175–1185

Srivastava R, Gupta JRP, Parthasarthy H (2010b) Comparison of PDE based and other techniques for speckle reduction from digitally reconstructed holographic images. Opt Lasers Eng 48(5):626–635

Srivastava R, Gupta JRP, Parthasarathy H (2011) Enhancement and restoration of microscopic images corrupted with poisson's noise using a nonlinear partial differential equation-based filter. Def Sci J 61(5):452

Srivastava S, Srivastava R, Sharma N, Singh SK, Sharma S (2012) A nonlinear complex diffusion based filter adapted to Rayleigh's speckle noise for de-speckling ultrasound images. Int J Biomed Eng Technol 10(2):101–117

Srivastava R (2011) A complex diffusion based nonlinear filter for speckle reduction from optical coherence tomography (OCT) images. In: Proceedings of the 2011 international conference on communication, computing & security, pp 259–264

Srivastava R, Gupta JRP, Parthasarathy H (2009) Complex diffusion based speckle reduction from digital images. In: 2009 proceeding of international conference on methods and models in computer science (ICM2CS). IEEE, pp 1–6

Srivastava R, Gupta JRP (2010) A PDE-based nonlinear filter adapted to Rayleigh's speckle noise for de-speckling 2D ultrasound images. In: International conference on contemporary computing. Springer, Berlin, Heidelberg, pp 1–12

Takezawa K (2005) Introduction to nonparametric regression, vol 606. John Wiley & Sons

Voci F, Eiho S, Sugimoto N, Sekibuchi H (2004) Estimating the gradient in the Perona-Malik equation. IEEE Signal Process Mag 21(3):39–65

Wahba G (1990) Spline models for observational data, vol 59. Siam

Whittaker ET (1922) On a new method of graduation. Proc Edinb Math Soc 41:63–75

You YL, Kaveh M (2000) Fourth-order partial differential equations for noise removal. IEEE Trans Image Process 9(10):1723–1730

Yu Y, Acton ST (2002) Speckle reducing anisotropic diffusion. IEEE Trans Image Process 11(11):1260–1270

Chapter 8
Computer-Aided Diabetic Retinopathy Diagnosis Using Conventional and Deep Learning Techniques—A Comparison

S. Valarmathi and R. Vijayabhanu

Abstract Diabetes mellitus, widely called diabetes, is a metabolic disorder that results in high glucose (blood sugar) level because of the inadequate production of the hormone called insulin. Diabetes, which affects the eyes and damages the retinal vessels, is known as diabetic retinopathy (DR). DR may even lead to blindness without prior symptoms. The existing handcrafted-based methods do not provide potential results in detecting retinal features and morphological differences in the retinal fundus images. Predominantly, in health care, deep learning makes a tremendous part in providing promising and more accurate results since it works well with unlabeled data and can easily learn low-level features from the training data. It acts as an assistant for the ophthalmologists in robustly examining retinal abnormalities. This chapter aims to discuss the state-of-the-art systems for detecting diabetic eye diseases with both traditional and deep learning techniques, and a statistical comparison is also made using various performance metrics. On comparing the accuracies and other metrics, the deep learning technique, convolutional neural network (CNN) solely surpasses the other conventional methods such as morphological methods, geometrical methods, pixel-based methods, and so on. Further, in this chapter, publicly available pertinent retinal image databases are also discussed.

Keywords Convolutional neural network (CNN) · Deep learning · Diabetes mellitus · Diabetic retinopathy

8.1 Introduction

Diabetes is one of the most commonly found disorders in a large number of people, which can cause damages across the body when the sugar level is uncontrolled or untreated. It can affect the vessels, kidneys, nerves, heart, eyes, and so on. According to global reports (Stitt et al. 2016), nearly 382 million individuals have been diagnosed with diabetes, and by 2030, it is predicted to attain 592 million. By early

S. Valarmathi (✉) · R. Vijayabhanu
Department of Computer Science, Avinashilingam Institute for Home Science and Higher Education for Women, Coimbatore, Tamil Nadu, India

© The Author(s), under exclusive license to Springer Nature Singapore Pte Ltd. 2022
N. Kumar et al. (eds.), *Advance Concepts of Image Processing and Pattern Recognition*,
Transactions on Computer Systems and Networks,
https://doi.org/10.1007/978-981-16-9324-3_8

disease detection, about 90% of people with diabetes can be extricated and treated appropriately to avoid future consequences. Most patients having diabetes for over ten years are more susceptible to the chance of getting diabetic retinopathy (DR). Diabetic retinopathy harms the retinal blood vessels, and it could even pave the way for blindness when it is left untreated.

8.2 Deep Learning

In this contemporary era, deep learning has become a robust tool as it is more powerful and being employed in various areas. Deep learning is an optimal technique to machine learning, and it is a multilayered model consisting of hierarchically structured architecture that can easily learn low and high-level features in the data, which can produce phenomenal results. Deep learning is widely known for five main popular reasons: (I) greater accuracy is achieved when the model is trained with an enormous volume of data; (II) computing hardware cost is reduced; (III) flexibility and efficiency; (IV) advances and development in the graphical processing units (GPUs) and its performance; (V) rapid transformation and advancement in machine learning algorithms. Thus far, standard multinational organizations such as Apple and Microsoft employ deep learning techniques to solve tedious tasks such as face detection, speech recognition, and so on. Deep learning follows a multilayered approach where the input data is processed in each of the layers, and the output is generated in an appropriate fashion. Generally, in the first layer, pixel extraction is performed in the input image data, which is followed by edge detection in the second layer. Other layers may perform segmentation and some other manipulation. The final layer performs a classification task which provides more accurate and potential results. Without human intervention, the retinal fundus image features can be learned using various powerful learning techniques in the deep learning model.

8.2.1 Deep Learning Applications

Over the years, deep learning has made a massive impact on several fields. In automated driving, deep learning has been used to detect objects like traffic lights, stop signs, and pedestrians, which automatically decreases accidents. In aerospace, satellite object detection and location can be found using deep learning, and it can also detect failures and faults in various complex systems by anomaly detection techniques. In automatic machine translation, deep learning has achieved greater results while translating both texts and images without using any preprocessing techniques. In automatic text generation, the deep learning model can easily learn to frame sentences, punctuate sentences, and it can even spell the texts. Some of the popular voice assistants developed by the standard organizations are Alexa, Cortana, and Siri by Amazon, Microsoft and Apple, respectively, which are conquering the modern world by its powerful and seamless user interactions and speech recognition. In agriculture, deep learning plays a vital role in automated irrigations, crop monitoring, disease detection, weed detection, and so on. In robotics, biomedical, and healthcare systems, deep learning techniques are employed to achieve greater and optimal results.

8.2.2 Deep Learning in Medical Image Processing

Nowadays, medical records are digitalized increasingly, and electronic health record (EHR) plays an integral role in analyzing a patient's history more effectively. Some of the conventional techniques used in medical image processing are support vector machine (SVM), K-nearest neighbors (KNN), neural network, and so on. On the other hand, some of the deep learning techniques are convolutional neural network (CNN), long short-term memory (LSTM), recurrent neural network (RNN), generative adversarial network (GAN), extreme learning, and so on. The various deep

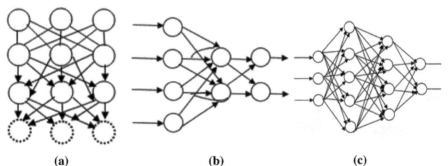

(a)   (b)   (c)

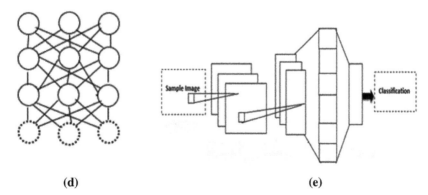

(d)   (e)

Fig. 8.1 **a** Deep belief network (DBN), **b** recurrent neural network (RNN), **c** deep neural network (DNN), **d** deep Boltzmann machine (DBM), **e** convolutional neural network (CNN)

Table 8.1 Comparison of deep learning architectures

Network architecture	Network description
1. Deep belief network (DBN) DBN is a kind of unsupervised probabilistic algorithm with multiple layers in which lower layers are directed from the above layers and the topmost layers are undirected (See Fig. 8.1)	**Pros**: Greedy-based learning strategy is used to train the model layer-by-layer which helps in achieving optimal results **Cons**: For large dataset, it is difficult to optimize the parameters
2. Recurrent neural network (RNN) RNN is a kind of neural network where the current layer's input can be fetched from the previous layer's output. The hidden layer is one of the most important features of RNN (See Fig. 8.1)	**Pros** 1. In time series prediction, RNN works well since it learns and remembers previous outputs 2. It provides greater results in character recognition, speech recognition, and other related tasks 3. LSTM, bidirectional long short-term memory (BLSTM) are some of the variations in RNN **Cons** 1. It is a tedious task to train an RNN model 2. While using activation functions such as ReLU and tanh, long sequences cannot be processed 3. Gradient vanishing is another issue in RNN
3. Deep neural network (DNN) For classification and regression purposes, DNN is mainly used and the network architecture consists of more than 2 layers (See Fig. 8.1)	**Pros** 1. DNN is one of the frequently used neural networks which has more than two layers with hidden layers **Cons** 1. For optimization and tuning the parameters, DNN models take enormous training time 2. Huge amount of labeled data is required to achieve greater and more accurate results
1. Deep Boltzmann machine (DBM) DBM is an undirected unsupervised graphical model having random hidden variables of multiple layers. It contains several hidden layers, unlike DBN (See Fig. 8.1)	**Pros**: It provides more robustness with the ambiguous data from the top down feedback network **Cons**: For large datasets, parameter optimization is a tedious and difficult task to perform in DBM
2. Convolutional neural network (CNN) CNN works well for two-dimensional (2D) data, and it has three layers in common: convolutional layer, pooling layer, and the fully connected layer (See Fig. 8.1)	**Pros** 1. CNN works well for image data and provides more accurate results 2. The CNN model can be trained fast, and it can learn both high- and low-level features from the data **Cons** 1. For accurate and effective classification, CNN needs large set of labeled data

learning network architectures are shown and described in Fig. 8.1 and Table 8.1, respectively.

8.2.3 Convolutional Neural Network (CNN)

Convolutional neural network (CNN) can be used in analyzing medical images, classifying an image, localizing, detecting, and segmenting essential information. In detecting various heart diseases, lung diseases, cancer diagnosis, tumor detection, and so on. CNN plays an integral role with its powerful and automatic feature extraction and classification. A CNN is a kind of neural network which has distinct layers, each layer performing specific tasks. Initially, the CNN model (See Fig. 8.2) starts with an input layer where the input image consists of raw pixels which equal the number of neurons present which is then forwarded to the next layers such as pooling layers, convolutional layers, and finally the output from the previous layers are given to the fully connected layer. According to the severity levels or by detection rate, the processed input data can be classified into different classes by the fully connected layer.

Convolutional layer

Convolution is normally an operation been performed on two functions. Based on the image position, the pixel values are taken as one of the function values. On the other hand, filter (kernel) is the second function represented as a list of numerical array values. On computation, the output can be given by the dot product of 2 functions. Then the filter is moved to the other positions in an image, which is called a stride

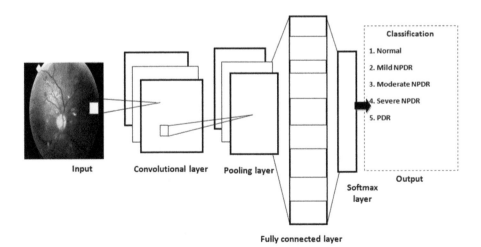

Fig. 8.2 CNN architecture for DR detection and classification

length. On processing the entire image, a feature map or activation map can be produced. The feature map is a kind of map generated which depends upon the filter.

For example, when an image of a tree is given as input in a CNN model, initially, the filter finds the low-level features in an image such as edges, lines, and so on. Then, progressively the high-level features are found, such as leaves, fruits, and so on, in the subsequent layers. In CNN, sparse connections can be established by connecting only the specific neurons to the next layer, where it learns only the meaningful and useful features and in turn the weights are reduced which improves the model accuracy. Parameter sharing can also be done, which reduces the memory storage.

Pooling layer

The pooling layer is mainly used to reduce image size and parameter count, which is called downsampling. Some of the pooling layers are average pooling, max pooling, L2-Normalization pooling, global pooling, and so on. Max pooling is the most commonly used pooling layer, within a filter it takes the greater region value, and the other values are dropped which effectively find the strongest activations.

Fully connected layer

Convolutional and pooling layers output are forwarded to the fully connected layer in which every neuron in the pooling layer is connected with the neurons in the fully connected layer. Using standard network model and learning methods with stochastic gradient descent and backpropagation technique improves the CNN performance and produces better results.

Softmax layer

A softmax layer is a final layer in a neural network in which multiple classes are involved, and the output will be produced in a probabilistic manner ranging from 0 to 1. It is a multi-class function which works well in classifying various class levels to produce effective and more accurate results.

8.2.3.1 CNN Training

The CNN model can be trained in two ways: training from scratch and transfer learning. Training the model from scratch can be a tedious and difficult task since it consumes much time in training, but in some scenarios, training the own model provides better results and accuracy when compared to pre-trained models.

Transfer learning

Transfer learning is one of the effective and common strategies which can be used to train a CNN model on a small amount of data since the model is already pre-trained with a large dataset. The ImageNet contest datasets and architectures are publicly available along with their weights and kernels. Transfer learning can be achieved with deep learning-based CNN architectures.

8.2.3.2 CNN Architectures

The ImageNet is a large database project mainly designed for object detection and visual recognition software-based research, which conducts an annual contest called ImageNet Large Scale Visual Recognition Challenge (ILSVRC). In this contest, various software programs compete with each other in accurately detecting and classifying the results. Some of the CNN architectures of top competitors in ILSVRC are discussed in Table 8.2.

8.3 Diabetic Eye Diseases

Diabetes Mellitus

Diabetes mellitus is a widely found disease where the pancreas secretes a small amount of insulin or no insulin, which results in insulin resistance. It affects people mostly aged between 45 and 64, and the actual diabetic causes are undefined or uncertain. Some of the risk factors involved are high blood pressure, obesity, unusual blood cholesterol level, polycystic ovary syndrome (PCOS), family history, and so on. Diabetes can be categorized into three types: Type I Diabetes, Type II Diabetes, and gestational diabetes. Over time, diabetic patients are more liable to get diabetic eye diseases. Diabetic retinopathy (DR) is stated to be the most commonly found diabetic eye disease, which leads to complete blindness. Some of the other diabetic eye diseases are diabetic macular edema, glaucoma, and cataracts.

Diabetic eye disease Symptoms

Mostly diabetic eye diseases may not show explicit symptoms at its earlier stage but when it progresses, few of the below symptoms may be addressed.

- Poor vision
- Light flashing
- Blurry vision
- Eye floaters
- Wavy eyesight.

8.3.1 Diabetic Retinopathy

Mostly diabetic patients suffer from diabetic retinopathy over time, and it is one of the greatest causes of vision loss. It mainly affects the retinal blood vessels such as arteries and veins, which act as a transportation medium in supplying nutrients to the retina. The normal retina and the DR-affected retina are shown in Figs. 8.3 and 8.4.

Table 8.2 Comparison of CNN architectures

CNN architecture	Developer and year	Description	No. of parameters
LeNet-5	Yann LeCun et al. in 1998	The LeNet-5 convolution network consists of 7-level which mainly is used in classifying digits and manual numbers written on cheques. To process high-resolution images, more convolutional layers are required	60,000 parameters used
AlexNet	Alex Krizhevsky et al. in 2012	The AlexNet architecture which outclassed all of the top competitors in ILSVRC ranked first place and achieved 15.3%, a top 5 error rate. It looks very similar to LeNet architecture but it was little deeper with more filters and convolutional layers. architecture consists of 11×11, 5×5, and 3×3 convolutions followed by max pooling, dropout, and ReLU activations	60 million parameters used
GoogLeNet	Szegedy et al. in 2014	In ILSVRC 2014 contest, GoogLeNet ranked first place and achieved a top 5 error rate of 6.67%. This GoogLeNet is also called as Inception V1which is mainly inspired by LeNet architecture and it used image distortions and batch normalization to decrease the number of parameters	4 million parameters used
VGGNet	Simonyan et al. in 2014	The VGGNet is the runner-up of ILSVRC 2014 contest which used 16 convolutional layers, and it is presently the most preferable CNN architecture	138 million parameters used

(continued)

Table 8.2 (continued)

CNN architecture	Developer and year	Description	No. of parameters
ResNet	Kaiming He et al. in 2015	ResNet achieves 3.57%, a top 5 error rate, and ranked 1st place with 152 layers. It introduces the feature called skip connections and batch normalization is also used	

Fig. 8.3 Normal retina

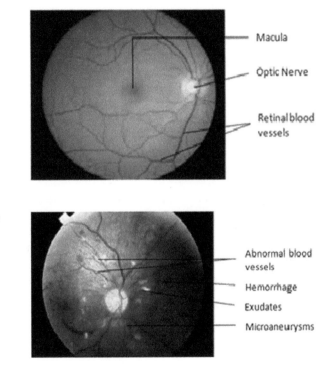

Fig. 8.4 DR-affected retina

8.3.1.1 Risk Factors

Diabetes—Duration period

Diabetes duration is one of the most vital and determining risk factors of diabetic retinopathy. Table 8.3 shows the DR prevalence percentage of both Type I and Type II diabetic patients. People with type 1 diabetes have a high risk of getting DR when compared to type 2 diabetic patients. Proliferative diabetic retinopathy (PDR) is highly found in type 1 patients, whereas type 2 diabetic patients mostly have

Table 8.3 Diabetes duration period

Duration	Type I diabetics (%)	Type II diabetics (%)
After 10 years	20	25
After 20 years	90	60
After 30 years	95	95

diabetic macular edema (DME). According to stats, female get affected than males in the ratio of 4:3.

Poor metabolic control and Heredity

Poor metabolic control is one of the risk factors which lead to DR progression and development in diabetic patients. DR is a kind of hereditary disease, which has a high effect on proliferative retinopathy.

Pregnancy and Hypertension

During the pregnancy period, the retinal features may change, and it gets accelerated. Hypertension can also increase or escalates the DR changes. Some of the other risk factors are anemia, obesity, and smoking.

8.3.1.2 Classification and Features

Diabetic retinopathy has been mainly classified into two types based on the morphological changes found in the retinal fundus images. Non-proliferative diabetic retinopathy (PDR) and proliferative diabetic retinopathy (PDR) are the types in which various severity levels are described. Diabetic maculopathy is closely related to PDR and NPDR because of their retinal effects. The macular region gets affected or swelled, and other retinal abnormalities can be found.

Non-proliferative Diabetic Retinopathy (NPDR)

Non-proliferative diabetic retinopathy (NPDR) is an early-stage diabetic retinopathy where four severity levels are listed below.

- Mild NPDR
- Moderate NPDR
- Severe NPDR
- Very severe NPDR.

NPDR Features

Microaneurysms are commonly found in the macular region or elsewhere in the retinal area formed mainly because of the focal dilation. It appears to leak fluids, lipids, and proteins formation and red spots around the macular area.

Table 8.4 ETDRS severity levels

S. No.	NPDR severity level	NPDR ophthalmoscopic features/symptoms
1	Mild NPDR	Minimum one microaneurysm observed
2	Moderate NPDR	• Microaneurysms and hemorrhage in 2 quadrants • Mild intraretinal microvascular abnormalities found • Cotton wool spots and exudates are found
3	Severe NPDR	Any 1 following symptoms, • Venous beading in 2 quadrants • Intraretinal abnormalities in 1 quadrant • Microaneurysms and extensive hemorrhages in 4 quadrants
4	Very severe NPDR	Any 2 following symptoms, • Venous beading in 2 quadrants • Intraretinal abnormalities in 1 quadrant • Microaneurysms and extensive hemorrhages in 4 quadrants

Retinal hemorrhages such as dot and blot hemorrhages are more commonly found, and the flame-shaped hemorrhage called superficial hemorrhages occurs mainly due to the leakages from the capillaries.

Hard exudates appear to be yellowish waxy patches found commonly in the macular region due to the leakage of the lipoproteins.

Cotton wool spots appear to be whitish lesions present in the retinal area.

Intraretinal microvascular abnormalities consist of irregular lines (red) connecting venules with the arterioles.

Venous abnormalities result in dilation, beading, and looping occurs in the capillary area.

ETDRS Risk levels

According to the **Early Treatment Diabetic Retinopathy Study (ETDRS)**, the NPDR severity levels have been classified in Table 8.4.

Proliferative Diabetic Retinopathy (PDR)

More than 50% of diabetic patients may get proliferative diabetic retinopathy (PDR) nearly after 25 years of the disease onset. PDR is most commonly found in type I diabetic patients. Neovascularization happens due to the development of abnormal blood vessels at the optic disc and elsewhere. PDR is the advanced diabetic retinopathy stage where the new abnormal vessels may spread and proliferate into the vitreous and the vitreous detachment and hemorrhage and fibrovascular membrane formation.

PDR Risk levels

Diabetic retinopathy study (DRS) has classified the PDR into two stages based on their high-risk characteristics (HRC). The ophthalmoscopic features of the two PDR types are described in Table 8.5.

Table 8.5 PDR risk levels

S. No.	PDR severity level	PDR ophthalmoscopic features/symptoms
1	PDR without HRC/early PDR	Early neovascularization found at the optic disc or elsewhere
2	PDR with HRC	• Optic disc neovascularization less than 1/4 disc region with preretinal and vitreous hemorrhage • Optic disc neovascularization with region 1/4 to 1/3 with or without preretinal and vitreous hemorrhage • Optic disc neovascularization greater than 1/2 disc region with both preretinal and vitreous hemorrhage

Fig. 8.5 Diabetic macular edema

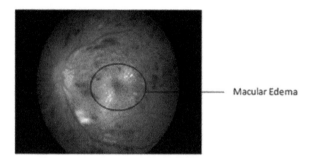

8.3.2 *Diabetic Macular Edema (DME)*

Diabetic macular edema as shown in Fig. 8.5 mainly affects diabetic patients due to the fluid which gets accumulated in the macular region of the retina. The macula, which is present at the center of the retina, gets swelled and thickened due to liquid formation which results in vision distortion.

DME Causes

DME occurs mainly due to the abnormal fluid leakage, fluid accumulation in the retinal macular region. Some of the other causes of DME are listed below.

• May develop after eye surgery
• Age-based macular degeneration
• Blood vessel blockage
• Inflammatory diseases.

8.3.3 *Glaucoma*

Glaucoma is one of the commonly found eye diseases which damage the optic nerve when the intraocular pressure is increased. It is shown in Fig. 8.6 and occurs because of the high pressure, which is accumulated inside the eye. It leads to complete vision

Fig. 8.6 Glaucoma

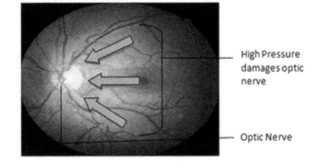

High Pressure damages optic nerve

Optic Nerve

loss or total blindness without prior symptoms when left untreated. Proper medications can be followed to reduce or lower the eye pressure, which prevents vision loss.

Glaucoma Causes

Glaucoma is one of the hereditary diseases that can be caused mainly due to the blood vessel blockage. Aqueous humor, a kind of fluid, which flows out of the eye through a medium, when that medium gets blocked, the fluid accumulates, which in turn produces high eye pressure. Other few causes are serious eye infection, injury, and also due to inflammatory conditions.

Glaucoma Risk factors

Some of the risk factors associated with glaucoma are listed below.

- Poor vision
- People above 40 years
- High blood pressure
- High eye pressure
- Diabetes mellitus
- Heart disease
- Thin corneas
- Family history.

Glaucoma Types

The most important and commonly found glaucoma types are open-angle glaucoma and angle-closure glaucoma. Some of the other types which are rarely found are listed: secondary glaucoma, normal-tension glaucoma, and pigmentary glaucoma.

8.3.4   Cataracts

A cataract induces a cloudy vision in the eye lens, and it develops slowly when the eye proteins form clumps, which results in blurry or cloudy vision. The retina converts the light, which passes via lens into signals, and it is sent to the brain through optic nerves. Cataracts are slowly developed and most commonly found in older people than others as shown in Fig. 8.7.

Cataract Causes

Cataract can be caused by various reasons. Some of those are listed below.

- Diabetes mellitus
- Continuous smoking
- Using steroids for longer period
- Ultraviolet radiation
- Trauma.

Cataract Risk factors

The risk factors involved in cataracts are listed below.

- Obesity
- High blood pressure
- High sun exposure
- Family history
- High exposure to radiations
- Diabetes mellitus
- Older age.

Cataract Types

There are several types of cataracts, which affect various parts of the eyes. Some of those are presented based on the locations and developments: nuclear cataracts, cortical cataracts, posterior capsular cataracts, congenital cataracts, secondary cataracts, and so on.

Fig. 8.7 Cataracts

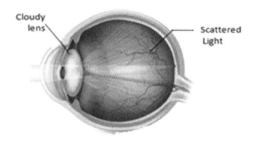

8.4 A Review on Retinal Image Databases

DRIVE—Digital Retinal Images for Vessel Extraction (DRIVE) is a database, which consists of retinal images that have been developed to make a comparative examination of blood vessel segmentation. Some of the morphological retinal blood vessel features are tortuosity, length, angles, and patterns for evaluating ophthalmologic and cardiovascular diseases like diabetes, neovascularization, and so on. The DRIVE image dataset consists of 400 diabetic records between the age group of 25–90 years, which was obtained in the Netherland from the DR Screening program. 40 images were selected randomly, and it was found that 33 of those do not possess any DR signs, whereas 7 of those showed early signs of mild DR. The images were acquired through a non-mydriatic 3CCD camera (Canon CR5).at 45° field of view (FOV).

STARE—STructured Analysis of the Retina (STARE) is a project started by Michael Goldbaum in 1975 which consisted of 40 fundus images of 605×700 size with 24 bits/resolution.

MESSIDOR—MESSIDOR is a research project mainly used to segment and detect lesions in a color fundus images in retinal ophthalmology. From three ophthalmology departments, 1200 color fundus images were acquired using a 3CCD camera using 8 bits/plane.

DIARETDB—DIARETDB is one of the standard diabetic retinopathy databases which contains 89 color images, out of which 84 show mid-NPDR signs, and the other 5 are normal images acquired by fundus camera with 50° FOV.

HRF—High-resolution fundus images database has been established to support studies made on automatic segmentation on retinal images. The segmentation database consists of 45 images, out of which 15 images are healthy, the other 15 images are DR-affected, and another 15 images are glaucomatous patient's data.

8.5 Diagnosing Diabetic Eye Diseases

Diabetic eye diseases can be diagnosed as early as possible to prevent from blurry vision or vision loss. Many computer-assisted diagnostic systems (CAD) have been developed for earlier disease prediction, which helps ophthalmologists in detecting retinal anatomic abnormalities.

8.5.1 Diagnosing Without Deep Learning Techniques

Diagnosing without using deep learning techniques can be explained using some of the retinal anatomic features for segmentation and classification purposes as shown in Fig. 8.8.

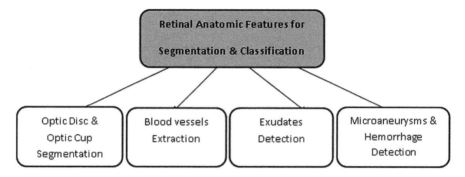

Fig. 8.8 Retinal anatomic features for segmentation and classification

8.5.1.1   Optic Disc and Optic Cup Segmentation

The optic disc (OD) is one of the primary anatomic features which detect the optic cup, fovea, and macular region in fundus images. The central axis and a contour of the optic disc can be fetched using segmentation and localization of the disc boundaries. Mostly geometrical and morphological-based methods are used in the literature for extracting the optic disc as described below.

Alshayeji et al. (2017) produced 95.91% of accuracy using gravitational law-based edge detection, and for optic disc segmentation, preprocessing techniques were applied. Abed et al. (2016) achieved an accuracy of 98.45%, which consumed less time than other techniques. One of the nature-inspired algorithms, swarm intelligence, was applied along with preprocessing. Nevertheless, the system was found too dependent on parameters and preprocessing techniques. In Díaz-Pernil et al. (2016), geometric-based methods have been used in CAD system named Hough circle cloud for optic disc localization, and it produced an accuracy of 99.6% while running on graphical processing units (GPUs) with small datasets.

The optic cup (OC) segmentation is also a monotonous task since the blood vessels are interlaced throughout the optic region. The pixel-based methods and the machine learning-based techniques are used in the literature for segmenting the optic cup, as described below.

Zilly et al. in (2017) fetched an accuracy of 94.1% while classifying small datasets using convolutional neural network (CNN) and entropy sampling for optic cup segmentation even though it failed to produce robust results when the sampling points are translated. Chakravarty and Sivaswamy (2017) produced 0.85 AUC value and the proposed system seems not much reliable or impressive in producing finer results and segmentation. They have created a combined model for optic disc and optic cup extraction. With coupled sparse dictionary techniques, the texture features were extracted and specifically applied for the optic cup detection. Arnay et al. in (2017) produced 0.79 AUC using ant colony optimization (ACO) for cup segmentation. However, it failed to produce results on some samples.

8.5.1.2 Blood Vessels Extraction

Detecting the structural variations in the retinal fundus images is such a critical and tedious task for clinicians. Through various CAD systems in the literature, accurate and more reliable results can be produced for the blood vessel segmentation. The literature techniques for the retinal blood vessel segmentation are separated into three categories: morphology-based techniques, supervised- and unsupervised-based techniques, which are described below.

Quellec et al. (2016) developed a model, which combines both normal features and abnormal features with contextualizing patient's data at proper adaptive granularity measures via data mining techniques for vessel segmentation. The system reported more potential and accurate results when compared to existing techniques on larger datasets, and it is found to be time-consuming and little expensive to implement. GeethaRamani and Balasubramanian (2016) used data mining techniques followed by preprocessing for blood vessel segmentation, which reported 95.36% of accuracy and achieved less specificity value with high processing time for blood vessel segmentation. Zhang et al. (2015) produced an accuracy of 95.05%, which used an unsupervised approach for blood vessel segmentation in which inconsistent labeling was optimized, and the text on dictionary was developed by furnishing key point descriptors to separate pixels from the blood vessel intensities. However, it produced false positive values throughout the disc and peripheral region, and it can be improved in the future by using various preprocessing techniques.

8.5.1.3 Exudates Detection

Exudate is one of the primary signs of diabetic retinopathy detection, which appears to be blob-shaped yellow-colored pigment, which presents on the retinal region. The literature techniques for the exudate detection are classified into three categories: machine learning-based techniques, mathematical morphology-based, and pixel-based techniques, as described below.

To distinguish retinal vessels from the exudate area, Imani and Pourreza (2016) used a morphological-based component-analysis model with 340 images. To distinguish the exudates from the usual retinal features, dynamic thresholding techniques and morphology-based methods are used, and it produced an AUC of 0.961. Prentasic and Lončarić (2016) used 50 images for exudate segmentation with a deep CNN technique. The optic disc segmentation was performed using probability maps, and the Frangi filter was used for blood vessel segmentation, which produced 0.78 F-score value. In Akram et al. (2014), used 1410 images for removing disc and blood vessels with mean–variance technique. For selecting exudate points and classifying between exudate and non-exudate pixels, bank filters and Gaussian-based integration in m-mediods model achieved an AUC score of 0.97.

8.5.1.4 Microaneurysms and Hemorrhage Detection

Microaneurysms and hemorrhage detection and extraction are a tedious and challenging task because of the variations in the features such as color, size, and textural properties in the fundus images. Various researchers are still working on this area to develop more reliable and potential lesion detection systems. The literature techniques for detecting microaneurysms and hemorrhage are discussed below using machine learning techniques, mathematical morphology-based techniques, supervised, and unsupervised-based techniques.

Habib et al. (2017) proposed a system, which detects microaneurysms in which Gaussian-based matching filter is used to extract specific features, and it is fed into tree-based ensemble classifier which classifies true and false points. The receiver operating curve (ROC) value was found to be 0.415, and the drawbacks found are overfitting, and the appropriate feature selection technique is undefined. Pereira et al. (2014) used kirsch, Gaussian, and median filters to the greener portion of a color image to extract microaneurysms features, and it is fed into multi-agent model to distinguish true microaneurysm pixels from the region. The ROC was found to be 0.24, and the drawback found here as it could not manipulate all optimal locations in a particular region. Agrawal et al. (2013) used various image processing techniques like feature extraction, preprocessing technique—contrast enhancement to provide a novel and improved microaneurysm detection. Adal et al. (2014) used a machine learning-based algorithm for microaneurysm detection followed by normalization and scale-based techniques for feature extraction. However, the authors obtained an AUC of 0.36 while using semi-supervised algorithms for classifying microaneurysm points.

Xiao et al. (2017) used a normalization technique to improve the image contrast and also to remove specific features which are fovea, disc, and vessels in a fundus image. For hemorrhage localization, Gaussian techniques were employed, and it yielded a sensitivity of 93.3% and specificity value of 88%. Lahmiri and Shmuel (2017) proposed a detection system which consists of three steps with 108 images. Based on the intensity variations, the image localization was performed, and the brightest features were extracted, and then the classifier was used to separate the hemorrhage pixels from the background image. The accuracy was found to be 100% in detecting the hemorrhage. Zhou et al. (2016) proposed a system in which before hemorrhage detection, Otsu's segmentation technique was used to subdue the vascular structure. The performance metrics reported are 92.6% sensitivity value, specificity value of 94% with 219 images.

8.5.1.5 DR Diagnosis Systems

Many DR-based diagnostic systems have been successfully developed in the literature, which produces effective results in detecting various stages of diabetic retinopathy. The primary goal of DR-based CAD systems is to separate the retinal anatomic features from the lesions. Few DR-based diagnostic systems in the literature are discussed below.

In Al-Bander et al. (2018), proposed a CAD system for optic disc segmentation and fovea with Kaggle and MESSIDOR datasets, which consists of 11,200 images. A multi-scale deep learning model was used for the fovea, and optic disc segmentation and the accuracy reported were 97% and 96.7% on the MESSIDOR dataset and 96.6% and 95.6% on Kaggle datasets. Dash and Bhoi (2017), proposed a tool for vascular structure segmentation, which produced 0.955 accuracy and 0.954 on CHASE_DB1 and DRIVE datasets of 58 images. Vo and Verma (2016) proposed a model based on deep learning, EyePACS, and MESSIDOR datasets of 91,402 images were used. The system used two deep networks, and hybrid-based color features are extracted to produce 0.891 and 0.887 ROC values. However, the above-discussed computer-based diagnostic systems are less impressive and applicable to produce finer results while classifying various stages of diabetic retinopathy.

8.5.2 Diagnosing with Deep Learning Techniques

Diabetic retinopathy is one of the lethal diseases which must be diagnosed early in order to prevent its prevalence. Most computerized tools seem to be effective in early DR detection, and employing deep learning techniques ultimately improves the prediction accuracy. One of the powerful deep learning techniques, convolutional neural network (CNN), has provided very impressive and potential results that outclassed human inspection, especially in medical imaging and healthcare systems. CNN is a feed-forward network, and it is one of the dominant deep learning approaches which have three layers in common, convolutional layer, pooling layer, and fully connected layers. Each CNN layer consists of numerous neurons where the input data is processed in each layer, and it is combined, overlapped to produce better image details. This process is carried out in all the CNN layers until accurate results are achieved. The CNN-based literature techniques for detecting diabetic retinopathy are discussed below.

Peng et al. (2019) proposed a deep learning model, DeepSeeNet is used, which detects the severity level of the age-based macular degeneration using bilateral CFP, and it achieved an accuracy of 0.967. Zhang et al. (2019) developed an automatic DR detection and grading system known as DeepDR which used transfer and ensemble learning such as ResNet 50 and Inception V3 which achieved 97.5% sensitivity value, 97.7% specificity value, and 97.7% AUC value. Saha et al. (2018) used a deep CNN, transfer-based learning such as AlexNet was used to grade the DR images as accept or reject classes. The work has been carried out using 7000 fundus images acquired from the EYEPACS by California Healthcare Foundation achieved an overall accuracy, sensitivity, and specificity of 100%. Arunkumar and Karthigaikumar (2017) used a deep learning-based feature extraction technique for diagnosing DR, age-based macular degeneration, retinal detachment, and a few more retinal diseases. Deep learning technique, deep belief network, was used along with multi-class SVM and obtained 96% accuracy, 79% sensitivity, and 97% specificity value.

8.6 Statistical Comparisons on with and Without Using Deep Learning Techniques

A statistical comparison can be made by comparing the performances of both with and without using deep learning techniques. Table 8.6 shows the result for optic

Table 8.6 Statistical comparison on with and without using deep learning techniques

1 Optic disc segmentation	2 Optic disc localization	3 Microaneurysms detection
1.1 DRIVE dataset	**2.1 DIARETDB1 dataset**	**3.1 ROC dataset**
Traditional Study: Tjandrasa et al. in (2012) **Technique**: Active contour **Accuracy**: 75.56% **Deep learning Study**: Tan et al. in (2017) **Technique**: Single state CNN **Accuracy**: 92.68%	**Traditional Study**: Sinha and Babu in (2012) **Technique**: One–one minimization **Accuracy**: 100% **Deep learning Study**: Alghamdi et al. in (2016) **Technique**: CNN **Accuracy**: 98.88%	**Traditional Study**: Javidi et al. in (2017) **Technique**: Sparse and dictionary learning **AUC**: 0.27 **Deep learning Study**: Haloi in (2015) **Technique**: Deep neural network **AUC**: 0.98
1.2 MESSIDOR dataset	**2.2 MESSIDOR dataset**	**3.2 MESSIDOR dataset**
Traditional Study: Aquino et al. in (2010) **Technique**: Morphological features **Accuracy**: 86% **Deep learning Study**: Lim et al. in (2015) **Technique**: CNN **Accuracy**: 96.40%	**Traditional Study**: Aquino et al. in (2010) **Technique**: Morphological features **Accuracy**: 99% **Deep learning Study**: Zhang et al. in (2018) **Technique**: R-CNN **Accuracy**: 99.9%	**Traditional Study**: Antal and Hajdu in (2012) **Technique**: Ensemble method **Accuracy**: 90% **Deep learning Study**: Haloi in (2015) **Technique**: DCNN **Accuracy**: 95.4%

disc segmentation, optic disc localization, and microaneurysms detection on various datasets.

Inferences

Overall, on comparing the performance and analyzing optic disc segmentation, optic disc localization, blood vessel segmentation, and DR-based lesion detection results, deep learning-based methods mainly CNN outperform the conventional methods as shown in Table 8.6. However, the deep learning-based methods are not time effective, and it is less robust, which also suffers from overfitting. So, to overcome all these drawbacks, more research work to be done in this particular area.

8.7   Future Directions

The deep learning-based techniques were found to achieve greater and more potential results for diagnosing diabetic retinopathy. Even though it suffers from various issues, data annotation is one among them where it demands ophthalmologist's services to label the color fundus images. To overcome this issue, a CNN model can be built with deep layers, and powerful learning algorithms can be developed to train a deep model on small datasets. Another challenging issue is a class imbalance in the datasets; usually, DR datasets contain very less DR-affected data and data with particular symptoms such as lesions, exudates, and so on than the normal ones. So, while training with such images, classification results in biased prediction, and class-specific solution. So, to overcome this drawback, data augmentation techniques other than geometrical transformations can be developed to balance the class. The deep learning model must provide better results when trained and tested on various datasets, but most of the time, it fails to provide effective results on different datasets, which is not robust. So, robustness must be improved using various deep learning methods so as to provide satisfactory performance over different datasets.

8.8   Conclusion

Diabetic retinopathy (DR) is one of the diabetic eye diseases most commonly found in diabetic patients, which even leads to vision loss without early symptoms. Many conventional DR diagnosis systems, which have been developed in the literature, do not provide satisfactory results on different datasets, whereas, with the advent of deep learning techniques, DR diagnosis systems are made powerful and more robust than conventional systems. Initially, deep learning, deep learning applications, deep learning in medical image processing, deep learning architectures, CNN architectures are discussed. Then, diabetic eye diseases are described along with their risk levels, and a review is done on retinal image databases. The DR diagnosis systems with and without deep learning techniques are discussed, followed by the statistical comparison made on traditional and deep learning techniques. Finally, the comparison states that deep learning-based technique; convolutional neural network outclasses the traditional methods to provide better and more accurate results.

References

Abed SE, Al-Roomi SA, Al-Shayeji M (2016) Effective optic disc detection method based on swarm intelligence techniques and novel pre-processing steps. Appl Soft Comput 49:146–163

Adal KM, Sidibé D, Ali S, Chaum E, Karnowski TP, Mériaudeau F (2014) Automated detection of microaneurysms using scale-adapted blob analysis and semi-supervised learning. Comput Methods Progr Biomed 114:1–10

Agrawal A, Bhatnagar C, Jalal AS (2013) A survey on automated microaneurysm detection in diabetic retinopathy retinal images. In: Proceedings of the international conference on information systems and computer networks (ISCON), pp 24–29, Mathura, India

Akram MU, Khalid S, Tariq A, Khan SA, Azam F (2014) Detection and classification of retinal lesions for grading of diabetic retinopathy. Comput Biol Med 45:161–171

Al-Bander B, Al-Nuaimy W, Williams BM, Zheng Y (2018) Multiscale sequential convolutional neural networks for simultaneous detection of fovea and optic disc. Biomed Signal Process Control 40:91–101

Alghamdi HS, Tang HL, Waheeb SA, Peto T (2016) Automatic optic disc abnormality detection in fundus images: a deep learning approach

Alshayeji M, Al-Roomi SA, Abed SE (2017) Optic disc detection in retinal fundus images using gravitational law-based edge detection. Med Biol Eng Comput 55:935–948

Antal B, Hajdu A (2012) An ensemble-based system for microaneurysm detection and diabetic retinopathy grading. IEEE Trans Biomed Eng 59(6):1720–1726

Aquino A, Geg undez-Arias ME, Marin D (2010) Detecting the optic disc boundary in digital fundus images using morphological, edge detection, and feature extraction techniques. IEEE Trans Med Imaging 29(11):1860–1869

Arnay R, Fumero F, Sigut J (2017) Ant colony optimization-based method for optic cup segmentation in retinal images. Appl Soft Comput 52:409–417

Arunkumar R, Karthigaikumar P (2017) Multi-retinal disease classification by reduced deep learning features. Neural Comput Appl 28:329–334

Chakravarty A, Sivaswamy J (2017) Joint optic disc and cup boundary extraction from monocular fundus images. Comput Methods Progr Biomed 147:51–61

Dash J, Bhoi N (2017) A thresholding based technique to extract retinal blood vessels from fundus images. Future Comput Inf J 2:103–109

Díaz-Pernil D, Fondón I, Peña-Cantillana F, Gutiérrez-Naranjo MA (2016) Fully automatized parallel segmentation of the optic disc in retinal fundus images. Pattern Recognit Lett 83:99–107

GeethaRamani R, Balasubramanian L (2016) Retinal blood vessel segmentation employing image processing and data mining techniques for computerized retinal image analysis. Biocybernññ Biomed Eng 36:102–118

Habib MM, Welikala RA, Hoppe A, Owen CG, Rudnicka AR, Barman SA (2017) Detection of microaneurysms in retinal images using an ensemble classifier. Inf Med Unlocked 9:44–57

Haloi M (2015) Improved microaneurysm detection using deep neural net. arXiv:1505.04424

Imani E, Pourreza HR (2016) A novel method for retinal exudate segmentation using signal separation algorithm. Comput Methods Progr Biomed 133:195–205

Javidi M, Pourreza HR, Harati A (2017) Vessel segmentation and microaneurysm detection using discriminative dictionary learning and sparse representation. Comput Methods Progr Biomed 139:93–108

Lahmiri S, Shmuel A (2017) Variational mode decomposition based approach for accurate classification of color fundus images with hemorrhages. Opt Laser Technol 96:243–248

Lim G, Cheng Y, Hsu W, Lee ML (2015) Integrated optic disc and cup segmentation with deep learning. 162–169

Peng Y, Dharssi S, Chen Q, Keenan TD, Agrón E, Wong WT, Chew EY, Lu Z (2019) DeepSeeNet: a deep learning model for automated classification of patient-based age-related macular degeneration severity from color fundus photographs. Ophthalmology 126:565–575

Pereira C, Vega D, Mahdjoub J, Guessoum Z, Gonçalves L, Ferreira M, Monteiro J (2014) Using a multi-agent system approach for microaneurysm detection in fundus images. Artif Intell Med 60:179–188

Prentasic P, Lončarić S (2016) Detection of exudates in fundus photographs using deep neural networks and anatomical landmark detection fusion. Comput Methods Progr Biomed 137:281–292

Quellec G, Lamard M, Erginay A, Chabouis A, Massin P, Cochener B, Cazuguel G (2016) Automatic detection of referral patients due to retinal pathologies through data mining. Med Image Anal 29:47–64

Saha SK, Fernando B, Cuadros J, Xiao D, Kanagasingam Y (2018) Automated quality assessment of colour fundus images for diabetic retinopathy screening in telemedicine. J Digit Imaging 31:1–10

Sinha N, Babu RV (2012) Optic disk localization using l 1 minimization. In: 19th IEEE international conference on in image processing (ICIP), pp 2829–2832

Stitt AW, Curtis TM, Chen M, Medina RJ, McKay GJ, Jenkins A, Lois N (2016) The progress in understanding and treatment of diabetic retinopathy. Prog Retin Eye Res 51:156–186

Tan JH, Acharya UR, Bhandary SV, Chua KC, Sivaprasad S (2017) Segmentation of optic disc, fovea and retinal vasculature using a single convolutional neural net. J Comput Sci 20:70–79

Tjandrasa H, Wijayanti A, Suciati N (2012) Optic nerve head segmentation using hough transform and active contours. Indonesian J Electr Eng Comp Sci 10(3):531–536

Vo HH, Verma A (2016) New deep neural nets for fine-grained diabetic retinopathy recognition on hybrid color space. In: Proceedings of the IEEE international symposium on multimedia (ISM), pp 209–215, San Jose, CA, USA

Xiao D, Yu S, Vignarajan J, An D, Tay-Kearney ML, Kanagasingam Y (2017) Retinal hemorrhage detection by rule-based and machine learning approach. In: Proceedings of the 39th annual international conference of the IEEE in engineering in medicine and biology society (EMBC), pp 660–663, Jeju Island, Korea

Zhang L, Fisher M, Wang W (2015) Retinal vessel segmentation using multi-scale textons derived from keypoints. Comput Med Imaging Gr 45:47–56

Zhang W, Zhong J, Yang S, Gao Z, Hu J, Chen Y, Yi Z (2019) Automated identification and grading system of diabetic retinopathy using deep neural networks. Knowl Based Syst 175:12–25

Zhang D, Zhu W, Zhao H, Shi F, Chen X (2018) Automatic localization and segmentation of optical disk based on faster R-CNN and level set in fundus image. In: Medical imaging 2018: image processing. international society for optics and photonics, vol 10574, p 105741U

Zhou L, Li P, Yu Q, Qiao Y, Yang J (2016) Automatic hemorrhage detection in color fundus images based on gradual removal of vascular branches. In: Proceedings of the IEEE international conference on image processing (ICIP), pp 399–403, Phoenix, AZ, USA

Zilly J, Buhmann JM, Mahapatra D (2017) Glaucoma detection using entropy sampling and ensemble learning for automatic optic cup and disc segmentation. Comput Med Imaging Gr 55:28–41

Chapter 9
Speckle Reduction in Ultrasound Images Using Hybridization of Wavelet-Based Novel Thresholding Approach with Guided Filter

Leena Jain and Palwinder Singh

Abstract The research work is aimed to explore and develop some new techniques in the area of speckle reduction. The algorithms are devised and implemented in transform domain for speckle reduction in ultrasound images. The high-quality ultrasound images can improve the accuracy and speed of subsequent image processing tasks and can yield better diagnosis results. In this work, we have proposed a hybrid method to reduce speckle from ultrasound images while maintaining a trade-off between speckle reduction and edge preservation. Extensive work is carried out by taking four different types of images for research which contains synthetic images, simulated images, noise-free ultrasound images, and real ultrasound images. The performance of proposed methods is measured by using different image quality metrics such as PSNR, MSE, COC, SSIM, FSIM, and EPI for all given sets of images except real ultrasound images. The real ultrasound images are compared by using visual comparison and metric SSI because it does not use reference images.

Keywords Ultrasound images · Speckle noise · Wavelet transform · Guided filter · Threshold method

9.1 Introduction

The ultrasound imaging utilizes high-frequency sound waves with a frequency of more than 20,000 Hz for the formation of ultrasound images (Chan and Perlas 2010). By transmitting these acoustic waves into human body through transducer, and receiving and processing the reflected acoustic waves the ultrasound images of internal body organs and tissues can be generated. When acoustic wave hits the body organ, it is reflected back then the transducer is switched off and listens to the ultrasound waves which are coming back from the body. These acoustic waves are converted to electric signals and send to the processor for developing an ultrasound

L. Jain (✉)
Global Group of Institutes, Amritsar, India

P. Singh
B.B.K D.A.V College for Women, Amritsar, India

© The Author(s), under exclusive license to Springer Nature Singapore Pte Ltd. 2022 155
N. Kumar et al. (eds.), *Advance Concepts of Image Processing and Pattern Recognition*,
Transactions on Computer Systems and Networks,
https://doi.org/10.1007/978-981-16-9324-3_9

image. Ultrasound imaging system is highly acceptable in veterinary and human diagnostics due to its non-invasive and versatile properties. Moreover, unlike the X-rays imaging system it does not use harmful ionizing radiations; therefore, the body organs or tissues under the examination are secure from harmful rays (Carovac et al. 2011).

Due to the fine quality of spatial and temporal resolution the accurate images of the human kidney, pancreas, gall bladder, breast, heart, and growth of fetus can be obtained using ultrasound imaging. Although ultrasound images have many advantages, it also suffers from severe artifacts like speckle noise and clutter (Ratliff 2013). The speckle noise is the major area of concern in ultrasound imaging. The problem which arises in ultrasound images due to speckle is the reduction of visible resolution and contrast. It adversely affects the interpretation result, and the extraction of evincive information becomes difficult. The aim of speckle reduction is to process an ultrasound image so that the processed image gives an accurate description using computer-aided diagnostics. In this chapter, Sect. 9.2 demonstrates the literature survey about speckle noise in ultrasound images and some well-known de-speckling techniques. The fundamental concepts of wavelet and noise reduction using wavelet thresholding are given in Sect. 9.3 of this chapter. The guided filter and proposed hybrid method using wavelet transform and guided filter is specified in Sect. 9.4 and Sect. 9.5, respectively. The experimental setup for research work and image quality metrics used for performance evaluation is discussed in Sects. 9.6 and 9.7, respectively. Sections 9.8–9.11 describe the results obtained for synthetic images, kidney phantom, cyst phantom, and real ultrasound images, respectively. In the end, the conclusion of the proposed work is given in Sect. 9.12 of this chapter.

9.2 Literature Survey

Speckle is a multiplicative noise which is inherited in all coherent imaging systems like radar, laser, and ultrasound. The two or more waves with the same frequency and constant phase difference are said to be from coherent sources. The majority of surfaces are very rough when scaled on optical wavelength. The coherent light when falls on a rough surface then reflected waves contains the contribution of many independent scattering areas. The interference of coherent and de-phased wavelets forms a granular pattern known as speckle (Joseph 1976). The speckle noise can be modeled as given below.

$$g(i, j) = f(i, j) * m(i, j) + n(i, j) \qquad (9.1)$$

where $f(i, j)$ and $g(i, j)$ are the given reference and degraded image, respectively. The given image is corrupted by additive noise $\eta(i, j)$ and multiplicative noise $m(i, j)$. The intensity of $\eta(i, j)$ is insignificant to $m(i, j)$, so we can rewrite mathematical equation of noise given in Eq. 9.1.

$$g(i, j) = f(i, j) * m(i, j) \qquad (9.2)$$

Speckle noise degrades the quality of ultrasound images by deteriorating edge definitions and significant features. It makes an area of interest difficult to analyze for radiologists and autonomous feature extraction (Wells and Halliwell 1981). In the last few decades, many different techniques have been proposed for de-speckling. The broad classification of these de-speckling methods is spatial domain and transform domain. The speckle noise reduction in the spatial domain includes direct manipulation of pixel values of an input image. The value of pixels is modified using some local neighborhood. The average intensity value of neighborhood pixels is calculated. This calculated average intensity can be used to replace the given pixel in arithmetic mean filter (Gonzalez and Woods 2008). The smallest neighborhood in spatial domain is 3×3 and can be raised upto a finite limit. The robust and adaptable versions of the mean filter based on statistical properties of the image were developed in Lee (1980); Frost et al. 1982), and (Kuan et al. 1985). A median filter is the nonlinear filter in which the value of a pixel in the filtered image was the median value of all intensity. The adaptive median filter was designed in such a way that the filter behavior changes when statistical properties in the given ultrasound image change. Adaptive filters for speckle reduction were proposed using local image statistics and geometrical ratio detectors based on local mean in image, stick orientation, local variance, and non-local mean in Loupas et al. (1989), Czerwinski et al. 1995, Qiu et al. (2004), and Guo et al. (2011), respectively. The spatial domain methods analyze the given image at a single scale, and these methods are not very effective for multiplicative noise. The partial differential equation-based nonlinear diffusion method was proposed by Perona and Malik (1990). The diffusion methods gained popularity because of their advantage of preserving edges and significant features while reducing noise. In (Choi and Jeong 2020), speckle reducing anisotropic diffusion was used for preprocessing and then wavelet transform was applied using symmetry characteristics. In (Sagheer and George 2017), a method based on weighted nuclear norm minimization (WNNM) for de-speckling from ultrasound images was proposed. The noise suppression in this approach was done in two phases, in which firstly homomorphic technique was applied followed by low-rank minimization. Secondly, the filtering was done using statistical properties of the ultrasound image. In (Ando et al. 2020), the image contrast was improved by reducing speckle noise using a fully convolution network. It was based on deep learning which contains an encoder, decoder, and final convolution layer to extract small feature maps by down-sampling an input image and then up-sampling to a get de-noised image. In (Chen et al. 2019), a novel method based on the super-pixel version of bilateral was proposed. The super-pixel segmentation has the advantages of sticking accurately to the boundaries and edges present in the given image.

9.3 Wavelet Thresholding

The transforms are used to obtain that information that is not readily available in raw form. The raw form of images is a spatial domain in which information about pixels, and their intensity is only available. But in order to get information about the frequency content of images, further processing of images or some transforms are required. The wavelet transform provides the ability to reduce noise from the given image by converting it into wavelet coefficients. The noise reduction from ultrasound images using wavelets is mainly based on the thresholding of wavelet coefficients. The wavelet coefficients are modified according to particular threshold value and thresholding rule. The wavelet transformation when applied on two-dimensional images the detail and approximation sub-bands are obtained. The process of decomposition is iterative, and choice of appropriate decomposition level is very crucial. The larger coefficients contain significant image data and coefficients close to zero or coefficients with smaller magnitude are considered as insignificant components of image. The de-noising using wavelet transformation involved four main key issues, which are:

- Selection of appropriate wavelet function/mother wavelet
- Selection of appropriate decomposition level
- Calculating accurate threshold value
- Selection of thresholding rule.

The choice of mother wavelet function must depend upon good localization characteristics in both the temporal and frequency domains. The proper trade-off between time and scale has also been taken into account while selecting the mother wavelet. The larger the decomposition level the narrower will be the frequency band which means the frequency resolution will be better. The ultrasound images are obtained in real time so in order to maintain a trade-off with time; we have not selected decomposition level beyond two. The value of the threshold should be optimal because small threshold value may not reduce noise and the resultant image will be still noisy and a large value of the threshold may produce an over-smoothed image because too many coefficients are set to zero. Many different techniques can be used for finding threshold values like VisuShrinkage (Donoho and Johnstone 1995), BayesShrinkage (Donoho and Johnstone 1994), Mod-BayesShrinkage (Chang et al. 2000), and NormalShrinkage (Elyasi and Zarmehi 2009). The researchers have also developed some hybrid methods using wavelet and fuzzy techniques (Kaur et al. 2002), wavelet, and bilateral filters (Wen and Qi 2015) for better speckle reduction and feature preservation. The hard and soft are the most commonly used thresholding rules (Balocco et al. 2010). A threshold value is calculated, and coefficients having value less than the threshold value are eliminated, and coefficients with value greater than thresholding value are kept unchanged.

$$\hat{f}_w = \begin{cases} f_w & \text{if } |f_w| > T \\ 0 & \text{if } |f_w| < T \end{cases} \tag{9.3}$$

In soft thresholding, if the value of coefficient is greater than the thresholding value, then the coefficients are reduced by using absolute value of threshold. The soft thresholding operator is defined in Eq. 9.4.

$$\hat{f}_w = \begin{cases} \text{sgn}(f_w)(|f_w| - T) & \text{if } |f_w| > T \\ 0 & \text{if } |f_w| < T \end{cases} \tag{9.4}$$

The hard thresholding function and soft thresholding functions have some limitations and scope of improvement. Due to 'keep or kill' property of hard thresholding, it makes images too much smooth. The hard thresholding function has no continuity at closed values, and reconstructed images may also contain distortion due to pseudo-Gibbs phenomenon. Moreover, the applicability of hard thresholding is also very limited as it processes only those coefficients which are smaller than the threshold. The soft thresholding rule has better continuity than hard thresholding rule but it may also produce over-smoothed images. A constant value compression can adversely affect the degree of approximation of reconstructed images. Therefore, the hard and soft thresholding functions have their own weaknesses. An improved thresholding function is proposed (Eq. 9.5), which has advantages of both hard and soft thresholding rules to overcome problems associated with traditional thresholding rules.

$$\hat{f}_w = \begin{cases} \text{sgn}(f_w)\left(|f_w| - \dfrac{T}{e^{\left(\frac{|f_w|-T}{T}\right)}}\right) & \text{if } |f_w| > T \\ 0 & \text{if } |f_w| < T \end{cases} \tag{9.5}$$

where f_w and $\hat{f}_w$ are wavelet coefficient and thresholded wavelet coefficients, respectively. The calculated threshold value and its own value can be used to adjust the proposed thresholding rule. If we use threshold value 'T' $= 0$ in the proposed thresholding rule, then it works like a hard thresholding rule, whereas for 'T' $= 1$, it becomes a soft thresholding rule. The formula given in Eq. 9.5 has also been used for the noise reduction from time series (Jain and Singh 2020).

9.4 Guided Filter

The guided image is modeled as a local linear model between guidance 'G' and filtering output 'q'. The image is filtered by using a guidance image, which may be the input noisy image itself or another image. The guided filter has good noise reduction and edge-preserving properties. The guided filter is known to be the fastest edge-preserving filter and also has the advantage of filtering close to edges (Sang et al. 2009). Suppose G is a guidance image, p and q are input and output images, respectively. Initially, a general translation variant filter is defined by including input, guidance, and output images. The processed output pixel can be obtained by using

Eq. 9.6.

$$q_i = \sum_j S_{i,j}(G) p_j \tag{9.6}$$

where i,j are pixel indices and $S_{i,j}$ is the filter kernel.

The guided image is modeled as a local linear model between guidance G and filtering output q. The following is the linear transform of 'G' in a window w_k which is centered at pixel k.

$$q_i = a_k G_i + b_k, \forall i \in w_k \tag{9.7}$$

where a_k, b_k are some constant linear coefficients in window w_k and to find linear coefficients the cost function in window w_k is defined as Eq. 9.8.

$$E(a_k, b_k) = \sum (a_k G_i + b_k - p_i)^2 + \varepsilon a_k^2 \tag{9.8}$$

where ε is the regularization parameter which prevents a_k from getting too large. The linear coefficients a_k and b_k can be determined using Eqs. 9.9 and 9.10, respectively.

$$a_k = \frac{\frac{1}{|w|} \sum_{i \in w_k} G_i p_i - \mu_k \overline{p}_k}{\sigma_k^2 + \varepsilon} \tag{9.9}$$

$$b_k = \overline{p}_k - \mu_k a_k \tag{9.10}$$

where μ_k and σ_k are the mean and standard deviation of 'G' in w_k, $|w|$ is the number of pixels in w_k and $\frac{1}{|w|} \sum_{i \in w_k} p_k$ is the mean of pi in w_k. So after computing a_k, b_k for all w_k in the image, the output can be computed according to formula given in Eq. 9.11.

$$q_i = \frac{1}{|w|} \sum_{k=i \in w_k} a_k l_i + b_k$$

$$= \overline{a}_i G_i + b_i \tag{9.11}$$

where $\overline{a}_i = \frac{1}{|w|} \sum_{k \in w_i} a_k$ and $\overline{b}_i = \frac{1}{|w|} \sum_{k \in w_i} b_k$.

The following algorithm can be used as a guided filter in which p is the input image, G is the guidance image, ε is the regularization factor, r is the radius of window, and output image q is obtained. The process of obtaining output image q is given below.

Step 1: Calculate initial statistical values

$\text{mean}_G = \text{Mean of guidance image}$

$mean_p$ = Mean of input image
$corr_G$ = Correlation of (G.*G)
$corr_{Gp}$ = Correlation of (G.*p)

Step 2: Calculate variance and covariances

ar_G = $corr_G$ – $mean_G$.*$mean_G$
$covar_{Gp}$ = $corr_{Gp}$ – $mean_G$.*$mean_p$
a = $covar_{Gp}$./($var_G + \varepsilon$)
b = $mean_p$ – a.*$mean_G$
$mean_a$ = Mean of 'a'
$mean_b$ = Mean of 'b'

Step 3: Calculate final output image

q = $mean_a$. *G + $mean_b$

where 'corr' is correlation, 'var' is variance, and 'covar' is covariance.

9.5   Proposed Hybrid Method

The speckle reduction based on wavelet thresholding decomposes the given image into approximation and detail sub-bands. The main focus in existing methods was on adjusting detail sub-band while keeping approximation sub-band unaltered. This is the reason that the approximation sub-band which represents the low-frequency component may still contain speckle noise. Moreover, the wavelet thresholding using a proposed thresholding rule gave good results for the low and medium ranges of speckle noise, but needs further improvement for high variance speckle noise (Adamo et al. 2013). Similarly, edges and features are also needed to preserve while denoising when the noise level is high. So, due to the inherited weakness of wavelet thresholding, which keeps approximation coefficients unchanged, a hybrid method is proposed. The proposed hybrid method uses wavelet thresholding in conjunction with the Guided filter to reduce speckle noise from ultrasound images. The guided filter is used to filter low-frequency LL sub-band wavelet coefficients. The de-noised ultrasound image constructed using inverse wavelet transform and comparative analysis performed with existing filters.

The performance comparison proved that the proposed method has further improved the noise reduction and edge preservation results for given ultrasound images. We have experimentally proved that the pre-filtration and post-filtration of ultrasound image with wavelet thresholding improves results of speckle reduction and edge preservation. Some authors have used bilateral filter instead of guided filter to reduce noise from low-frequency sub-band (He et al. 2013). Some other has also used the guided filter with wavelet thresholding, but they did not use the process of pre-filtration and post-filtration (Zhang et al. 2015).

The guided filter is used initially for de-speckling from the vital areas of ultrasound images. The guided filter is again applied to filter the low-frequency subband of wavelet coefficients for the given image. Adaptive wavelet shrinkage is applied to eliminate noisy detail coefficients. In the end, the post-filtration removes the remaining speckle noise from the reconstructed image. The detailed methodology of the proposed algorithm is given in Fig. 9.1.

The proposed hybrid algorithm for de-speckling ultrasound images is given as follows.

Step 1: Use guided filter for pre-filtration of ultrasound image corrupted by speckle noise.

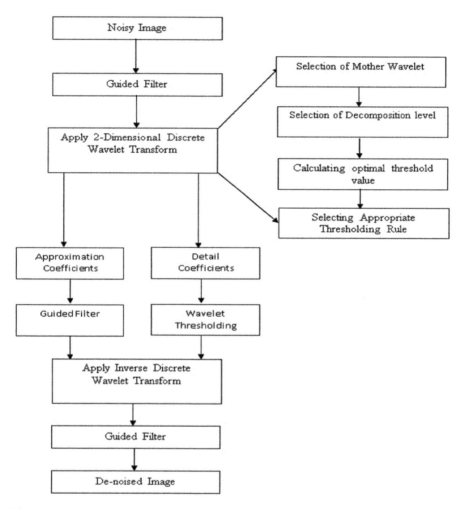

Fig. 9.1 Model of speckle reduction using proposed hybrid method

Step 2: Use DWT on noisy ultrasound image. The mother wavelet is taken as Daubechies (db8), and level 2 is used for decomposition.

Step 3: Use Eq. 9.12 for calculation of noise variance of given image.

$$\hat{\sigma}^2 = \left[\frac{\text{median}(|E_{i,j}|)}{0.675} \right]^2 \tag{9.12}$$

Step 4: Use Eq. 9.13 (VisuShrinkage) or Eq. 9.14 (BayesShrinkage) or Eq. 9.15 (Mod-BayesShrinkage) or Eq. 9.16 (NormalShrinkage) for the calculation of threshold value.

$$T_v = \sqrt{2 \log P \hat{\sigma}^2} \tag{9.13}$$

$$T_B = \frac{\hat{\sigma}^2}{\hat{\sigma}_y^2} \tag{9.14}$$

$$T_{NS} = C_{NS} \frac{\hat{\sigma}^2}{\hat{\sigma}_y^2} \tag{9.15}$$

$$T_{MB} = C_{MB} \frac{\hat{\sigma}^2}{\hat{\sigma}_y^2} \tag{9.16}$$

Step 5: The detail wavelet coefficients are treated with proposed thresholding rule (Eq. 9.5).

Step 6: Use the guided filter over approximation coefficients to remove noise from low-frequency sub-band.

Step 7: Use inverse wavelet transform on thresholded wavelet coefficients to obtain a final reconstructed image.

Step 8: Use guided filter to perform post-filtration of ultrasound image corrupted by speckle noise.

We have implemented the proposed hybrid method and compared it with existing speckle reduction methods. The comparison is made with traditional methods like Lee (1980), Frost et al. (1982), Kuan et al. (1985), Guided (Jain and Singh 2020), Bilateral (Zhang et al. 2015), SRAD (Zhang et al. 2016) and wavelet thresholding methods VisuShrinkage (Donoho and Johnstone 1995), BayesShrinkage (Donoho and Johnstone 1994), Modified-BayesShrinkage (Chang et al. 2000), Normal-Shrinkage (Elyasi and Zarmehi 2009) using soft (Balocco et al. 2010) and proposed thresholding rule (Adamo et al. 2013). The comparison is also made with some latest de-noising techniques hybrid-Zhang et al. (2015) and Neigh-Shrink-Sure (Yongjian and Acton 2002). The subjective evaluation along with objective metrics is used for visual evaluation which shows the eminence of proposed hybrid method over existing speckle reduction techniques.

9.6 Experimental Setup

The data set used for experimentation contains synthetic images, cyst, and kidney phantom, noise-free ultrasound images, and real ultrasound images. We have implemented 6 spatial domain, 13 transform domain, and 2 hybrid filters on 5 different data sets of images. The performance evaluation is done using 7 different image quality metrics.

9.6.1 Synthetic Images (Test Image-1)

In order to make quantitative and qualitative analyses of the proposed algorithm, the experiments are also conducted on the synthetic image given. The synthetic image of size 512*512 contains different geometric shapes with uniform regions and sharp edges. These images are used to check the effectiveness of different de-noising algorithms. The speckle noise is added in the synthetic image of variance between 0.02 and 0.06. The reference synthetic image and synthetic image corrupted with speckle noise of variance 0.04 are given in Figs. 9.2 and 9.3, respectively.

Fig. 9.2 Reference synthetic image

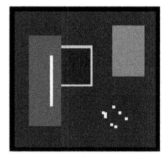

Fig. 9.3 Synthetic image corrupted with speckle noise of variance 0.04

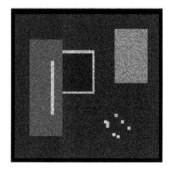

9.6.2   Kidney Phantom (Test Image-2) and Cyst Phantom (Test Image-3)

This field-II software simulates the transmission of ultrasound waves, operation of transducer, and medium where ultrasound waves travel. The ultrasound images of different types and different speckle patterns can be obtained by adjusting parameters like transducer frequency, sampling frequency, number of active elements, and speed of sound. The kidney phantoms and cyst phantoms are simulated using the filed-II simulator developed by Jensen (Chen et al. 2005). The time of simulation for different images is generally very large but it can be kept down by using low sampling frequency. The apodization, attenuation, and focusing of transducer are perfectly handled by different programs in simulator (Jensen and Svendsen 1992). In our research work, we have taken simulated images of cyst phantom and synthetic kidney obtained using the field-II simulator for experimentation and comparative analysis. The kidney phantom has been created by using 128 elements in the transducer, 64 active elements, and transducer frequency is adjusted to 7 MHz. The kidney phantom formation also begins with field.m which initializes the field system. The task of creating a file for scatterers in the phantom is done using the make_scatteres.m routine. The scatterer map of kidney is stored in file kidney_cut.bmp. The field simulation is performed using sim_kidney.m, and data is stored under rf_data folder as RF-files. Finally, kidney image is created by calling the routine make_polar.m. The kidney phantom and cyst phantom formed using field-II simulator are given in Figs. 9.4 and 9.5, respectively.

   The cyst phantom has been created by using 192 elements in transducer, 64 active elements, and transducer frequency is adjusted to 3.5 MHz. The program field.m is executed to initialize the field system, and the task of creating a file for scatterers in the phantom is done using themk_pht.m routine. The field simulation is performed using sim_img.m, and data is stored under the rf_data folder as RF-files. The data

Fig. 9.4 Kidney phantom created using Field-II

Fig. 9.5 Cyst phantom created using Field-II

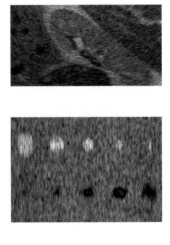

Fig. 9.6 Real ultrasound
images of pelvic kidney
right-side longitudinal
(Jensen 1996)

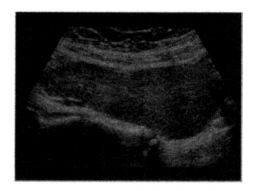

stored in given folder is then processed using the make_image.m routine, and a cyst
phantom image is formed.

9.6.3 Real Ultrasound Images (Test Image-4)

The de-noising and feature preserving capabilities of proposed and existing methods
are also evaluated using real ultrasound images of pelvic kidney on right-side longitudinal. The images (Fig. 9.6) are taken from the data set provided by Gelderse Vallei
hospital, Netherland, for research purpose (Jensen 1996).

9.7 Image Quality Metrics

The quality of an image can be measured by using objective and subjective image
quality metrics. In the applications in which evaluation is made visually with the
human eye, the subjective quality measures are used. But in some applications, the
image quality is evaluated automatically on the basis of automated mathematical
measures (Real Ultrasound Images Database 2016).

Image quality metric	Formula	Description
Mean Square Error (Wang et al. 2004)	$\text{MSE} = \frac{1}{MN} \sum_{i=1}^{M} \sum_{j=1}^{N} (\hat{f}(i,j) - f(i,j))^2$	$f(i,j)$ is the original image of size $M \times N$ $\hat{f}(i,j)$ is the de-speckled image
Peak Signal-to-Noise Ratio (Wang et al. 2004)	$\text{PSNR} = 10 \log \frac{I_M^2}{\text{MSE}}$	I_M is the maximum intensity available MSE is the mean square error

(continued)

(continued)

Image quality metric	Formula	Description
Structural Similarity Index (Wang and Bovik 2009)	$$SSIM = \frac{\left(2\mu_f\mu_{\hat{f}}+K_1\right)\cdot\left(2\sigma_{f\hat{f}}+K_2\right)}{\left(\mu_f^2+\mu_{\hat{f}}^2+K_1\right)\cdot\left(\sigma_f^2+\sigma_{\hat{f}}^2+K_2\right)}$$	K_1 and K_2 are stabilizing constants μ_f and $\mu_{\hat{f}}$ are mean intensity of original and de-noised image, respectively σ_f^2, $\sigma_{\hat{f}}^2$, and $\sigma_{f\hat{f}}$ are variance of reference image, variance of de-speckled image, and covariance, respectively
Coefficient of Correlation (Gonzalez and Woods 2008)	$$COC = $$ $$\frac{\sum(f(i,j)-\mu_f)\cdot(\hat{f}(i,j)-\mu_{\hat{f}})}{\sqrt{\sum(f(i,j)-\mu_f)^2\cdot\sum(\hat{f}(i,j)-\mu_{\hat{f}})^2}}$$	$f(i,j)$ is the original image of size $M \times N$ $\hat{f}(i,j)$ is the de-speckled image μ_f and $\mu_{\hat{f}}$ are mean intensity of original and de-noised image, respectively
Edge Preservation Index (Sattar et al. 1997)	$$EPI = $$ $$\frac{\sum_{i=1}^{M}\sum_{j=1}^{N}(\Delta f(i,j)-\Delta\mu_f)(\Delta\hat{f}(i,j)-\Delta\mu_{\hat{f}})}{\sum_{i=1}^{M}\sum_{j=1}^{N}(\Delta f(i,j)-\Delta\mu_f)^2(\Delta\hat{f}(i,j)-\Delta\mu_{\hat{f}})^2}$$	$\Delta f(i,j)$ and $\Delta\hat{f}(i,j)$ are obtained by applying high-pass filter on original and de-noised images $\Delta\mu_f$ and $\Delta\mu_{\hat{f}}$ are the average intensities of $\Delta f(i,j)$ and $\Delta\hat{f}(i,j)$, respectively
Feature Similarity Index (Lin et al. 2011)	$$FSIM = \frac{\sum_{k\in\Omega}SIM_L(k)PC_M(k)}{\sum_{k\in\Omega}PC_M(k)}$$	PC and G_f are phase congruencies and gradient magnitude, respectively $SIM_L(k) = [SIM_{PC}(k)]^\alpha[SIM_G(k)]^\beta$ $SIM_{PC}(k) = $ $\left[\frac{2PC_f(k)PC_{\hat{f}}(k)+T_1}{PC_f^2(k)PC_{\hat{f}}^2(k)+T_1}\right]$ $SIM_G(k) = $ $\left[\frac{2G_f(k)G_{\hat{f}}(k)+T_2}{G_f^2(k)G_{\hat{f}}^2(k)+T_2}\right]$ T_1 and T_2 are positive stabilizing constants $PC_M(k) = \max[PC_1(k), PC_2(k)]$

(continued)

(continued)

Image quality metric	Formula	Description
Speckle Suppression Index (Sheng and Xia 2004)	$SSI = \frac{\sigma_{\hat{f}}}{\mu_{\hat{f}}} \cdot \frac{\mu_f}{\sigma_f}$	μ_f and $\mu_{\hat{f}}$ are mean intensity of original and de-noised image, respectively $\sigma_f^2, \sigma_{\hat{f}}^2$ are variance of reference image and de-speckled image, respectively

9.8 Experiment Results and Discussions for Synthetic Images (Test Image-1)

The synthetic images are created in MATLAB, and speckle noise of variance 0.02 to 0.06 is artificially added in these images. Table 9.1 contains the PSNR values on Test Image-1 using different de-speckling methods. Figure 9.7 contains plot of PSNR

Table 9.1 PSNR values of Test Image-1 using spatial and transform domain methods

De-noising technique	Variance				
	0.02	0.03	0.04	0.05	0.06
Noisy Image	25.6002	23.8607	22.6353	21.6725	20.8915
Guided Filter	34.7384	33.6782	32.8053	32.0474	31.4635
Frost Filter	32.3181	29.6586	27.6822	26.2199	25.0226
Kuan Filter	28.1584	27.8574	27.7145	27.4871	27.1983
Lee Filter	30.2775	29.8065	29.4472	29.0721	28.7601
SRAD Filter	32.7256	29.7458	27.1573	26.7511	25.07
Bilateral Filter	33.2118	29.9255	27.3876	26.5474	24.9885
VisuShrinkage-soft	33.504	32.0648	31.1059	30.3567	29.769
VisuShrinkage-Proposed	34.5764	33.2567	32.0495	30.6085	29.7857
BayesShrinkage-Soft	28.7753	26.9343	25.814	24.9016	24.1531
BayesShrinkage-Proposed	29.094	26.98	25.9891	24.7561	24.1057
Mod-BayesShrinkage-Soft	34.4193	33.9006	32.9371	30.9977	29.4042
Mod-BayesShrink-Proposed	35.6785	34.2256	33.2551	31.3865	29.4839
NormalShrinkage-Soft	34.7512	33.5749	32.6417	31.8906	31.1991
NormalShrinkage-Proposed	35.3088	34.2143	33.1686	32.0617	31.1871
Neigh-Shrink-Sure	35.4993	33.9453	31.7618	30.0552	28.2982
Hybrid-Zhang	35.6253	34.0256	31.8223	30.2135	28.7632
Hybrid-Proposed	**35.8401**	**34.693**	**33.311**	**32.5642**	**31.3871**

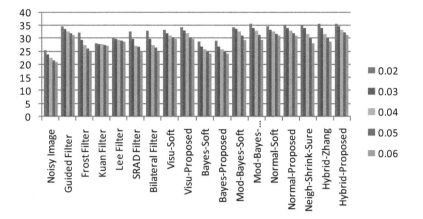

Fig. 9.7 Plot of PSNR versus noise variance using different de-noising methods for Test Image-1

obtained using different de-noising methods for Test Image-1. The various parameters of these de-noising techniques are carefully selected for efficient and optimal performance. The local window size as taken as 5×5 for spatial domain filters like Lee, Frost, Kuan, guided, bilateral, and diffusion filter. The degree of smoothness is taken as 650.2*36 for guided filter and bilateral filter. It is observed that for given parameters, the performance of the proposed hybrid method outperforms every existing speckle reduction technique. Table 9.2 contains the MSE values of different speckle reduction methods, and Fig. 9.8 contains plot of MSE using different de-noising methods for Test Image-1. The values and plot of SSIM obtained using different speckle reduction techniques are given in Table 9.3 and Fig. 9.9, respectively. The larger value of SSIM indicates the better structural similarity between the reference image and de-noised image. The values of COC and FSIM are shown in Tables 9.4 and 9.5, respectively. Similarly, plots of COC and FSIM obtained using different speckle reduction techniques are given in Figs. 9.10 and 9.11, respectively. Table 9.6 contains value of EPI obtained using different speckle reduction techniques. Figure 9.12 contains a plot of EPI obtained for Test Image-1 using different noise reduction methods. It is observed that the results of the proposed hybrid method outperform the existing de-noising methods.

The de-speckled Test Image-1 is given in Figs. 9.13 and 9.14, which indicates that the proposed method has better speckle reduction capabilities.

9.9 Experiment Results for Kidney Phantom (Test Image-2)

The PSNR, MSE, and SSIM values obtained for Test Image-2 using different de-noising techniques are given in Table 9.7. Test Image-3 is simulated using field-II simulator so these images already contain speckle noise. So, unlike Test Image-1, we

Table 9.2 MSE values of Test Image-1 using spatial and transform domain methods

De-noising technique	Variance				
	0.02	0.03	0.04	0.05	0.06
Noisy Image	179.0851	267.3089	354.4456	442.4141	529.5768
Guided Filter	38.1299	50.3422	70.883	95.271	102.5604
Frost Filter	61.8393	67.8782	74.0844	90.5828	106.4226
Kuan Filter	99.3665	106.4986	110.0596	115.9769	123.9521
Lee Filter	61	67.9872	73.8509	80.5143	86.5109
SRAD Filter	43.1257	44.3689	71.2914	76.3256	101.8733
Bilateral Filter	48.5546	54.2215	60.3254	75.9621	104.6168
VisuShrinkage-soft	39.019	50.4205	60.4074	69.8971	78.5771
VisuShrinkage-Proposed	32.4227	41.4014	52.2205	60.0155	67.4795
BayesShrinkage-Soft	113.6454	165.8244	214.5869	264.8045	314.611
BayesShrinkage-Proposed	103.6324	159.5528	211.9225	243.7053	313.9227
Mod-BayesShrinkage-Soft	36.8975	46.5301	53.0553	73.1113	98.2115
Mod-BayesShrink-Proposed	29.8761	38.1982	45.5959	61.1347	81.2664
NormalShrinkage-Soft	32.4133	41.9408	47.556	52.9687	58.1116
NormalShrinkage-Proposed	28.2896	34.2467	41.3488	47.7496	53.9914
Neigh-Shrink-Sure	27.3296	35.2152	43.3409	54.2037	86.2191
Hybrid-Zhang	27.2284	34.2569	40.4587	52.251	50.4655
Hybrid-Proposed	**27.1085**	**31.7114**	**39.8893**	**44.6273**	**50.3719 asis>**

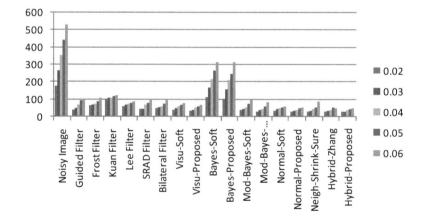

Fig. 9.8 Plot of MSE versus noise variance using different de-noising methods for Test Image-1

Table 9.3 SSIM values of Test Image-1 using spatial and transform domain methods

De-noising technique	Variance				
	0.02	0.03	0.04	0.05	0.06
Noisy Image	0.4313	0.3638	0.3217	0.2924	0.2711
Guided Filter	0.8526	0.8007	0.7927	0.778	0.7667
Frost Filter	0.8421	0.8212	0.8014	0.7837	0.767
Kuan Filter	0.7875	0.7692	0.7522	0.737	0.7118
Lee Filter	0.8026	0.7984	0.7844	0.7604	0.7553
SRAD Filter	0.8471	0.8162	0.781	0.7797	0.7662
Bilateral Filter	0.8459	0.8119	0.7794	0.7707	0.7259
VisuShrinkage-soft	0.8481	0.8388	0.8306	0.825	0.8193
VisuShrinkage-Proposed	0.8497	0.8405	0.8335	0.8279	0.8221
BayesShrinkage-Soft	0.7132	0.6523	0.6142	0.582	0.5517
BayesShrinkage-Proposed	0.7234	0.6612	0.6293	0.5866	0.5482
Mod-BayesShrinkage-Soft	0.8408	0.8171	0.8047	0.7944	0.7737
Mod-BayesShrink-Proposed	0.8729	0.8541	0.837	0.8112	0.8041
NormalShrinkage-Soft	0.8683	0.8475	0.8303	0.8299	0.8204
NormalShrinkage-Proposed	0.8771	0.8531	0.8315	0.8296	0.8178
Neigh-Shrink-Sure	0.8777	0.8565	0.8388	0.8148	0.8053
Hybrid-Zhang	0.8852	0.8577	0.8359	0.8220	0.8186
Hybrid-Proposed	**0.8917**	**0.8626**	**0.8567**	**0.8374**	**0.8216**

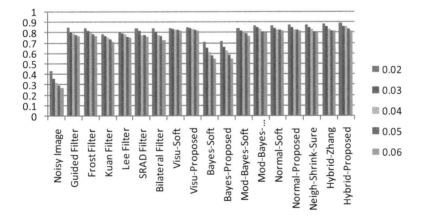

Fig. 9.9 Plot of SSIM versus noise variance using different de-noising methods for Test Image-1

Table 9.4 COC values of Test Image-1 using spatial and transform domain methods

De-noising technique	Variance				
	0.02	0.03	0.04	0.05	0.06
Noisy Image	0.9661	0.9502	0.935	0.9202	0.9061
Guided Filter	0.9926	0.9893	0.9883	0.9898	0.9834
Frost Filter	0.9901	0.9885	0.9869	0.9857	0.9826
Kuan Filter	0.9809	0.9796	0.9789	0.9778	0.9763
Lee Filter	0.9891	0.9879	0.9869	0.9857	0.9846
SRAD Filter	0.9912	0.9897	0.9863	0.9793	0.9725
Bilateral Filter	0.9904	0.9898	0.9852	0.9775	0.9710
VisuShrinkage-soft	0.985	0.983	0.9812	0.9796	0.9781
VisuShrinkage-Proposed	0.9871	0.9857	0.9844	0.983	0.9817
BayesShrinkage-Soft	0.9781	0.9682	0.9591	0.9498	0.9409
BayesShrinkage-Proposed	0.9876	0.9826	0.9774	0.972	0.9675
Mod-BayesShrinkage-Soft	0.9909	0.987	0.9838	0.9799	0.9769
Mod-BayesShrink-Proposed	0.9956	0.9938	0.9918	0.9898	0.9874
NormalShrinkage-Soft	0.9946	0.9929	0.9921	0.9906	**0.989**
NormalShrinkage-Proposed	0.9947	0.9931	0.9914	0.9898	0.9873
Neigh-Shrink-Sure	0.9961	0.9951	0.9917	0.9875	0.9812
Hybrid-Zhang	0.9959	0.9948	0.992	0.9901	0.9876
Hybrid-Proposed	**0.9967**	**0.9957**	**0.9931**	**0.991**	**0.989**

have not added any speckle noise artificially using MATLAB. The results of PSNR and MSE for the proposed hybrid method are better than traditional speckle reduction techniques. The SSIM shows that the results of proposed hybrid, Hybrid-Zhang, and NormalShrinkage-Soft are similar. Table 9.8 contains FSIM, COC, and EPI values obtained using different de-noising techniques. The FSIM shows significant improvement in feature similarity using proposed hybrid method. The results of COC are similar for proposed hybrid and Neigh-Shrink-Sure. Similarly, EPI value for proposed hybrid method is almost the same as that of NormalShrinkage-Soft. Figures 9.15, 9.16, 9.17, 9.18, 9.19, and 9.20 contain plots of PSNR, MSE, SSIM, FSIM, COC, and EPI for Test Image-3. Although the results for given metrics are not showing any negative trend, still there is a scope of improvement.

The subjective analysis of Test Image-2 obtained after de-noising by different de-noising techniques is given in Figs. 9.21 and 9.22. The results of proposed hybrid method show visual improvement in de-noised images when compared with existing speckle reduction methods.

Table 9.5 FSIM values of Test Image-1 using spatial and transform domain methods

De-noising technique	Variance				
	0.02	0.03	0.04	0.05	0.06
Noisy Image	0.9599	0.9452	0.9317	0.9193	0.9089
Guided Filter	0.9712	0.9601	0.9487	0.9377	0.9289
Frost Filter	0.9696	0.9587	0.9483	0.9395	0.9317
Kuan Filter	0.9618	0.9465	0.9381	0.9283	0.9212
Lee Filter	0.9694	0.9623	0.9553	0.9486	0.9423
SRAD Filter	0.9711	0.9587	0.9432	0.9359	0.9237
Bilateral Filter	0.9679	0.9412	0.9296	0.9238	0.9141
VisuShrinkage-soft	0.9667	0.9548	0.9501	0.9456	0.9415
VisuShrinkage-Proposed	0.9765	0.9624	0.951	0.9423	0.9347
BayesShrinkage-Soft	0.9636	0.9561	0.9455	0.937	0.9283
BayesShrinkage-Proposed	0.9677	0.9564	0.9461	0.9363	0.9283
Mod-BayesShrinkage-Soft	0.9754	0.9624	0.9508	0.9406	0.9319
Mod-BayesShrink-Proposed	0.9794	0.968	0.9541	0.9415	0.9320
NormalShrinkage-Soft	0.9783	0.9663	0.9569	0.9484	0.9408
NormalShrinkage-Proposed	0.9793	0.9674	0.9569	0.9464	0.9389
Neigh-Shrink-Sure	0.9755	0.9627	0.9506	0.9403	0.9302
Hybrid-Zhang	0.9791	0.968	0.9576	0.9462	0.9392
Hybrid-Proposed	**0.9796**	**0.9682**	**0.9576**	**0.9492**	**0.9411**

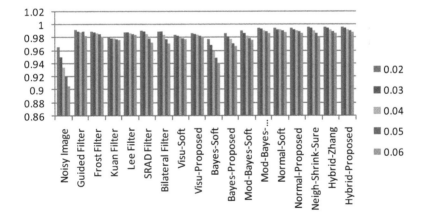

Fig. 9.10 Plot of COC versus noise variance using different de-noising methods for Test Image-1

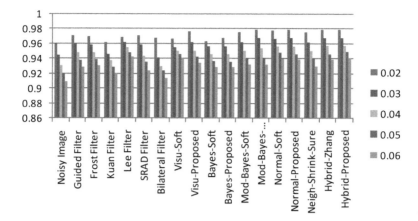

Fig. 9.11 Plot of FSIM versus noise variance using different de-noising methods for Test Image-1

Table 9.6 EPI values of Test Image-1 using spatial and transform domain methods

De-noising technique	Variance				
	0.02	0.03	0.04	0.05	0.06
Noisy Image	0.3664	0.3015	0.2611	0.2356	0.2116
Guided Filter	0.8853	0.8589	0.8434	0.7851	0.7363
Frost Filter	0.8526	0.814	0.7757	0.7277	0.692
Kuan Filter	0.6821	0.6435	0.6398	0.6346	0.6266
Lee Filter	0.7281	0.6373	0.6084	0.5792	0.521
SRAD Filter	0.8923	0.8567	0.8523	0.7936	0.7453
Bilateral Filter	0.8258	0.802	0.7764	0.7143	0.6944
VisuShrinkage-soft	0.6475	0.6367	0.6299	0.6282	0.6229
VisuShrinkage-Proposed	0.6544	0.6483	0.6424	0.6357	0.6226
BayesShrinkage-Soft	0.4596	0.3894	0.344	0.3099	0.2921
BayesShrinkage-Proposed	0.4923	0.4179	0.3682	0.3206	0.2915
Mod-BayesShrinkage-Soft	0.8552	0.8244	0.7826	0.7534	0.727
Mod-BayesShrink-Proposed	0.9248	0.9016	0.8626	0.8275	0.8128
NormalShrinkage-Soft	0.899	0.8929	0.8881	**0.8778**	0.8653
NormalShrinkage-Proposed	0.9131	0.9027	0.8862	0.8744	0.8643
Neigh-Shrink-Sure	0.9428	0.9368	0.8873	0.8676	0.8657
Hybrid-Zhang	0.9435	0.9364	**0.8936**	0.8773	**0.8674**
Hybrid-Proposed	**0.9462**	**0.9378**	**0.8936**	0.8771	**0.8674**

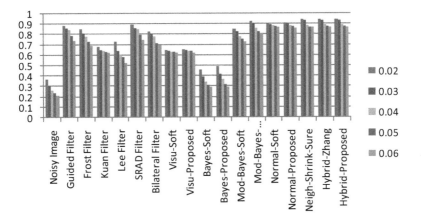

Fig. 9.12 Plot of EPI versus noise variance using different de-noising methods for Test Image-1

9.10 Experiment Results and Discussions for Cyst Phantom (Test Image-4)

Table 9.9 contains PSNR, MSE, and SSIM values obtained for Test Image-3 using different de-noising methods. The PSNR, MSE values for the proposed hybrid method exhibit significant improvement over existing speckle reduction methods. The SSIM values of normal shrink using soft thresholding are slightly better than the hybrid method. The values of FSIM, COC, and EPI are given in Table 9.10 for Test Image-3. The plots of PSNR, MSE, SSIM, FSIM, COC, and EPI obtained for Test Image-3 using different speckle reduction methods are given in Figs. 9.23, 9.24, 9.25, 9.26, 9.27, and 9.28, respectively. The FSIM and EPI values have been improved for the proposed hybrid method, whereas COC values of proposed hybrid method are same as that of Hybrid-Zhang method.

The subjective results are given in Figs. 9.29 and 9.30 of proposed hybrid method which shows visual improvement in de-noised images when compared with existing speckle reduction methods.

9.11 Experiment Results and Discussions for Real Ultrasound Images (Test Image-4)

The effectiveness of proposed hybrid and existing speckle reduction techniques is measured using SSI and visually, for ultrasound images. The original noise-free image is not available for real ultrasound images so metrics like PSNR, MSE, SSIM, COC, FSIM, and EPI cannot be used for performance comparison. Figs. 9.31 and 9.32 present the subjective evaluation for proposed hybrid filter and different existing speckle reduction methods for real ultrasound images. It is found that image filtered

Fig. 9.13 Results of Test
Image-1 wavelet
thresholding techniques

(a) VisuShrinkage-soft

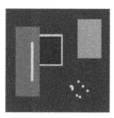

(b) VisuShrinkage-Proposed

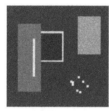

**(c) BayesShrinkage-
Soft**

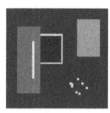

(d) Bayes-Proposed

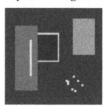

**(e) Mod-
BayesShrinkage-Soft**

(f) Mod-BayesShrink-Proposed

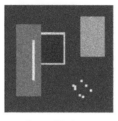

**(g) NormalShrinkage-
Soft**

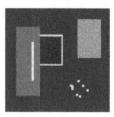

**(h) NormalShrinkage-
Proposed**

with proposed hybrid method shows visual improvement when compared against other speckle reduction methods. The minor visual differences are very difficult to check with naked eye; therefore, we have also used SSI as image quality metric. The performance comparison of SSI (Table 9.11) shows the eminence of proposed hybrid methods over existing speckle reduction techniques.

The results of proposed hybrid method are 5.66, 5.29, 5.82, 3.3, and 2.52% better than spatial domain techniques such as Lee (1980), Frost et al. (1982), Kuan et al. (1985), Bilateral (Zhang et al. 2015), and Guided filter (Jain and

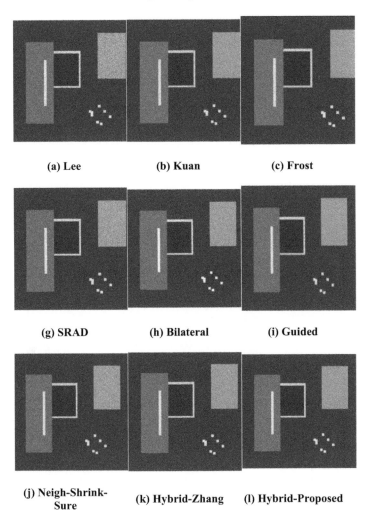

Fig. 9.14   Results of Test Image-1 using different de-noising techniques

Singh 2020), respectively, in terms of SSI. The SSI for proposed method is
3.57% better than diffusion method SRAD (Zhang et al. 2016). Similarly, proposed
hybrid method has given 2.15%, 2.09%, 0.97%, 0.94% improved SSI results than
VisuShrinkage (Donoho and Johnstone 1995), BayesShrinkage (Donoho and Johnstone 1994), Mod-BayesShrinkage (Chang et al. 2000), and NormalShrinkage (Elyasi
and Zarmehi 2009), respectively, for soft thresholding rule and 1.87%, 2.21%, 0.15%,
0.63% improved SSI results than VisuShrinkage (Donoho and Johnstone 1995),
BayesShrinkage (Donoho and Johnstone 1994), Mod-BayesShrinkage (Chang et al.
2000) and NormalShrinkage (Elyasi and Zarmehi 2009), respectively, for proposed

Table 9.7 PSNR, MSE, and SSIM values obtained for Test Image-2 using spatial and transform domain methods

De-noising technique	Image quality metrics		
	PSNR	MSE	SSIM
Noisy Image	15.8613	1686.341	0.1497
Guided Filter	16.7284	1359.0733	0.2161
Frost Filter	16.664	1368.2206	0.2065
Kuan Filter	16.4442	1371.7229	0.2142
Lee Filter	16.2588	1422.3616	0.1946
SRAD Filter	16.6348	1373.5216	0.2087
Bilateral Filter	16.2267	1463.8468	0.1988
VisuShrinkage-soft	16.1354	1537.5425	0.2043
VisuShrinkage-Proposed	16.2627	1509.7742	0.2043
BayesShrinkage-Soft	15.981	1614.6669	0.1848
BayesShrinkage-Proposed	16.0362	1529.1156	0.1843
Mod-BayesShrinkage-Soft	16.3054	1501.3184	0.2164
Mod-BayesShrink-Proposed	16.7352	1358.6863	0.2171
NormalShrinkage-Soft	16.3939	1358.7596	**0.2189**
NormalShrinkage-Proposed	16.4336	1359.0408	0.2187
Neigh-Shrink-Sure	16.7711	1367.6464	0.2185
Hybrid-Zhang	16.8316	1336.3542	**0.2189**
Hybrid-Proposed	**16.8812**	**1333.4139**	**0.2189**

thresholding rule. The proposed hybrid method gives 0.67%, 0.62% better results than Neigh-Shrink-Sure and Hybrid-Zhang, respectively, in terms of SSI.

Table 9.12 contains the execution time of different thresholding methods for all data set of images. It is observed that the proposed method is computationally expensive and takes more time as compared to existing speckle reduction methods. Although we have studied that the better de-noising results increase the speed of subsequent image processing tasks; however, the ultrasound images are obtained in real time, so execution time should also be reduced.

9.12 Conclusion

In concluding remarks, the improvement of SSI using real ultrasound images for the proposed hybrid method can be discussed here. The research work is aimed to explore and develop some new techniques in the area of speckle reduction from ultrasound images. The proposed hybrid method is showing remarkable improvement in edge preservation and speckle reduction. The results of the proposed hybrid method are 5.66%, 5.29%, 5.82%, 3.3%, and 2.52% better than spatial domain techniques such as Lee, Kuan, Frost, bilateral, and guided filter, respectively, in terms

Table 9.8 FSIM, COC, and EPI values obtained for Test Image-2 using spatial and transform domain methods

De-noising technique	Image quality metrics		
	FSIM	COC	EPI
Noisy Image	0.5917	0.5919	0.145
Guided Filter	0.6492	0.664	0.3027
Frost Filter	0.6511	0.65	0.2989
Kuan Filter	0.6357	0.6262	0.2722
Lee Filter	0.6346	0.6346	0.2876
SRAD Filter	0.6536	0.6115	0.2797
Bilateral Filter	0.6485	0.6596	0.264
VisuShrinkage-soft	0.6232	0.6416	0.2535
VisuShrinkage-Proposed	0.6238	0.6436	0.2511
BayesShrinkage-Soft	0.5989	0.6089	0.1982
BayesShrinkage-Proposed	0.5981	0.6174	0.2003
Mod-BayesShrinkage-Soft	0.6473	0.6538	0.2964
Mod-BayesShrink-Proposed	0.6686	0.6678	0.2962
NormalShrinkage-Soft	0.6508	0.6562	**0.3177**
NormalShrinkage-Proposed	0.6501	0.6597	0.3166
Neigh-Shrink-Sure	0.6619	**0.6712**	0.3155
Hybrid-Zhang	0.6682	0.6683	0.3163
Hybrid-Proposed	**0.6696**	**0.6712**	**0.3177**

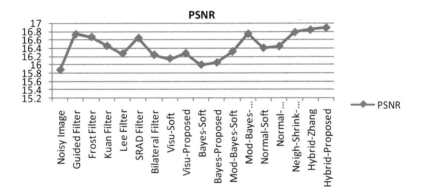

Fig. 9.15 Plot of PSNR using different de-noising methods for Test Image-2

of SSI. The SSI for the proposed method is 3.57% better than the diffusion method SRAD. Similarly, proposed hybrid method has given 2.15%, 2.09%, 0.97%, 0.94% improved SSI results than VisuShrinkage, BayesShrinkage, Mod-BayesShrinkage, and NormalShrinkage, respectively, for soft thresholding rule and 1.87%, 2.21%,

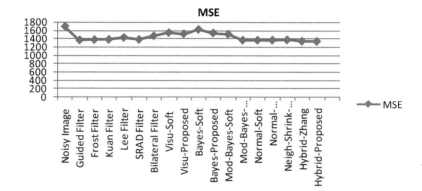

Fig. 9.16 Plot of MSE using different de-noising methods for Test Image-2

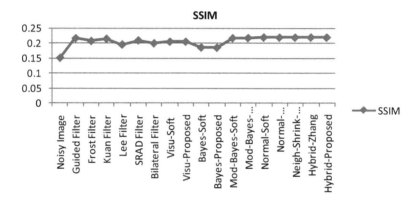

Fig. 9.17 Plot of SSIM using different de-noising methods for Test Image-2

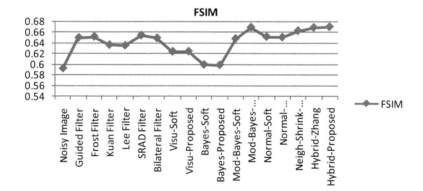

Fig. 9.18 Plot of FSIM using different de-noising methods for Test Image-2

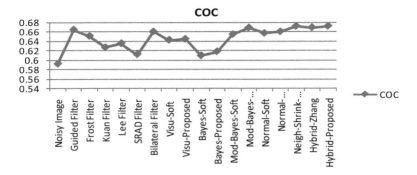

Fig. 9.19 Plot of COC using different de-noising methods for Test Image-2

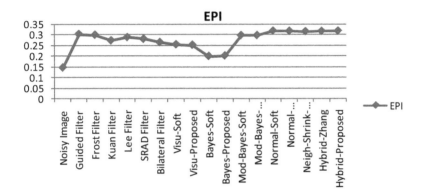

Fig. 9.20 Plot of EPI using different de-noising methods for Test Image-2

0.15%, 0.63% improved SSI results than VisuShrinkage, BayesShrinkage, Mod-BayesShrinkage, and NormalShrinkage, respectively, for proposed thresholding rule. The proposed hybrid method gives 0.67%, 0.62% better results than Neigh-Shrink-Sure, and Hybrid-Zhang, respectively, in terms of SSI. The proposed thresholding rule when compared against existing speckle reduction methods, it is observed that the proposed thresholding rule outperforms every existing method discussed. The proposed method shows better results in edge preservation and speckle reduction.

Fig. 9.21 Results of Test
Image-1 wavelet
thresholding techniques

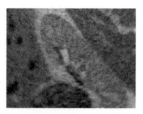

(a) VisuShrinkage-soft   (c) VisuShrinkage-Proposed

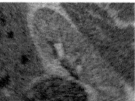

(d) BayesShrinkage-Soft   (f) Bayes-Proposed

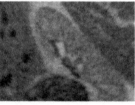

(g) Mod-BayesShrinkage-Soft   (i) Mod-BayesShrink-Proposed

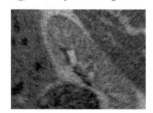
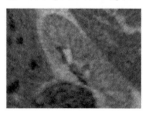

(j) NormalShrinkage-Soft   (l) NormalShrinkage-Proposed

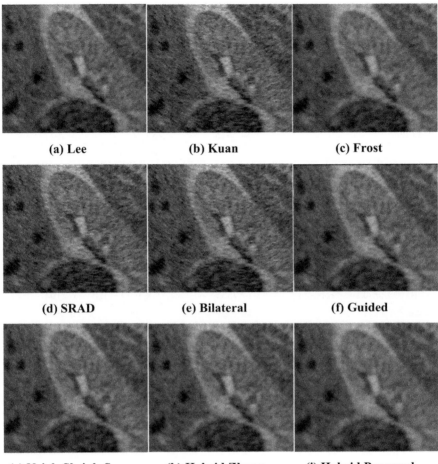

(a) Lee   (b) Kuan   (c) Frost

(d) SRAD   (e) Bilateral   (f) Guided

(g) Neigh-Shrink-Sure   (h) Hybrid Zhang   (i) Hybrid Proposed

Fig. 9.22 Results of Test Image-2 using different de-noising techniques

Table 9.9 Performance comparison of PSNR, MSE, and SSIM values obtained for Test Image-3 using different de-noising techniques

De-noising technique	Image quality metrics		
	PSNR	MSE	SSIM
Noisy Image	14.2077	2467.7957	0.2049
Guided Filter	14.7223	2192.0647	0.2192
Frost Filter	14.7455	2180.3476	0.2146
Kuan Filter	14.5415	2285.2424	0.2031
Lee Filter	14.5446	2181.1756	0.2187
SRAD Filter	14.7369	2130.0359	0.218
Bilateral Filter	14.6551	2226.2474	0.2175
VisuShrinkage-Soft	14.6972	2154.5632	0.2188
VisuShrinkage-Proposed	14.7082	2097.1234	0.2155
BayesShrinkage-Soft	14.743	2120.6098	0.2097
BayesShrinkage-Proposed	14.7436	2101.3222	0.2098
Mod-BayesShrinkage-Soft	14.7532	2061.4125	0.2194
Mod-BayesShrinkage-Proposed	14.7555	2058.3627	0.2194
NormalShrinkage-Soft	14.7569	2059.6905	**0.2196**
NormalShrinkage-Proposed	14.7572	**2056.5156**	0.2192
Neigh-Shrink-Sure	14.5661	2078.3249	0.2178
Hybrid-Zhang	14.7023	2074.3682	0.2195
Hybrid-Proposed	**14.7701**	**2056.5156**	0.2195

Table 9.10 Performance comparison of FSIM, COC, and EPI values obtained for Test Image-3 using different de-noising techniques

De-noising technique	Image quality metrics		
	FSIM	COC	EPI
Noisy Image	0.3518	0.1865	0.4507
Guided Filter	0.4318	0.3197	0.6114
Frost Filter	0.4226	0.2219	0.519
Kuan Filter	0.4241	0.222	0.4945
Lee Filter	0.4051	0.2828	0.5105
SRAD Filter	0.4307	0.321	0.6173
Bilateral Filter	0.4244	0.3095	0.5917
VisuShrinkage-soft	0.4158	0.3146	0.5596
VisuShrinkage-Proposed	0.4163	0.3106	0.5589
BayesShrinkage-Soft	0.393	0.3111	0.5787
Bayes-Proposed	0.4029	0.3163	0.5728
Mod-BayesShrinkage-Soft	0.4235	0.3201	0.6106
Mod-BayesShrink-Proposed	0.4328	0.3297	0.6212
NormalShrinkage-Soft	0.4338	0.3218	0.6201
NormalShrinkage-Proposed	0.4329	0.3232	0.6180
Neigh-Shrink-Sure	0.4332	0.3069	0.6179
Hybrid-Zhang	0.4329	**0.33**	6218
Hybrid-Proposed	**0.4344**	**0.33**	0.6227

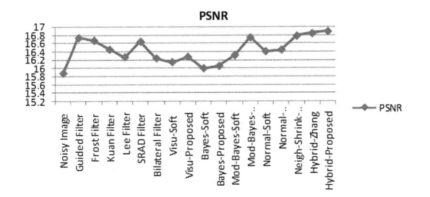

Fig. 9.23 Plot of PSNR using different de-noising methods for Test Image-2

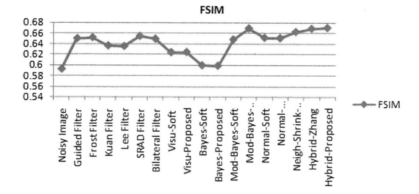

Fig. 9.24 Plot of MSE using different de-noising methods for Test Image-2

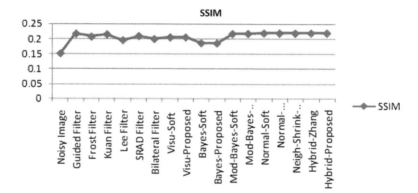

Fig. 9.25 Plot of SSLM using different de-noising methods for Test Image-2

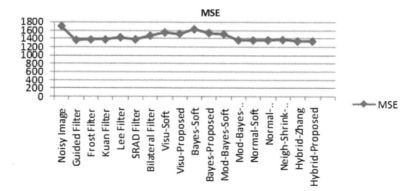

Fig. 9.26 Plot of FSIM using different de-noising methods for Test Image-2

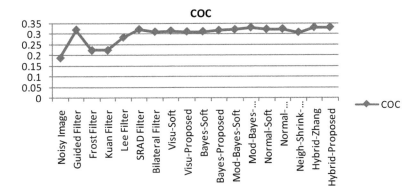

Fig. 9.27 Plot of COC using different de-noising methods for Test Image-3

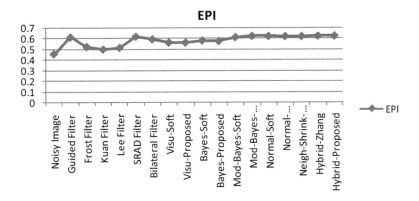

Fig. 9.28 Plot of EPI using different de-noising methods for Test Image-3

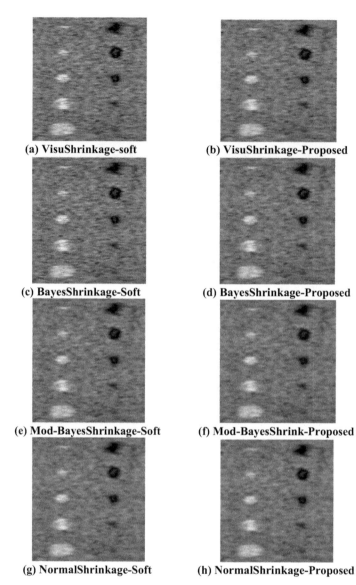

(a) VisuShrinkage-soft   **(b) VisuShrinkage-Proposed**

(c) BayesShrinkage-Soft   **(d) BayesShrinkage-Proposed**

(e) Mod-BayesShrinkage-Soft   **(f) Mod-BayesShrink-Proposed**

(g) NormalShrinkage-Soft   **(h) NormalShrinkage-Proposed**

Fig. 9.29 Results of Test Image-9.3 Wavelet thresholding techniques

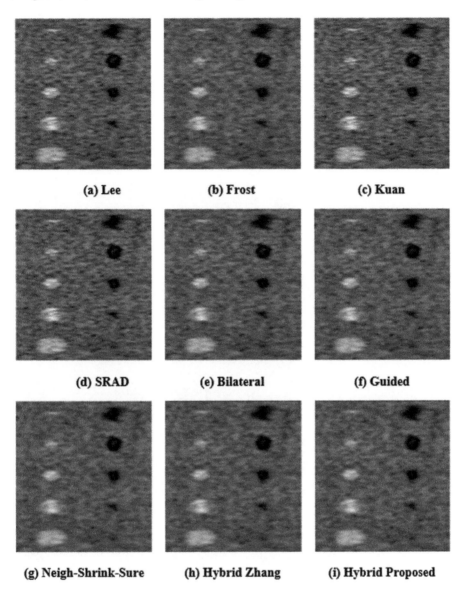

Fig. 9.30   Results of Test Image-3 using different de-noising techniques

Fig. 9.31 Results of Test
Image-4 wavelet
thresholding techniques

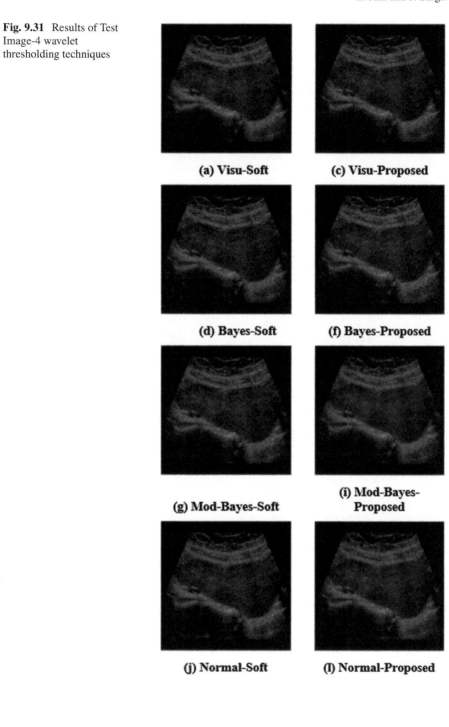

(a) Visu-Soft   (c) Visu-Proposed

(d) Bayes-Soft   (f) Bayes-Proposed

(g) Mod-Bayes-Soft   (i) Mod-Bayes-
                                 Proposed

(j) Normal-Soft   (l) Normal-Proposed

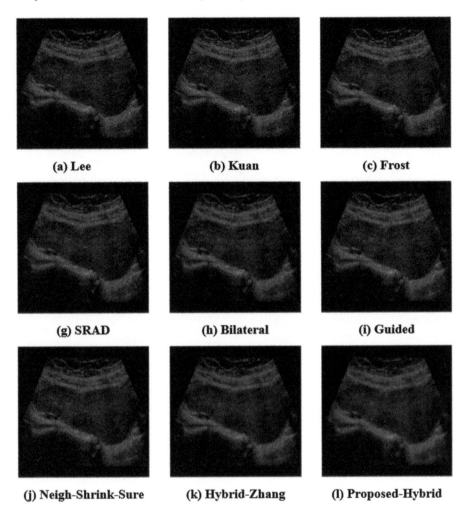

Fig. 9.32 Results of Test Image-4 using different de-noising techniques

Table 9.11 SSI values obtained for Test Image-4 using spatial and transform methods

Algorithm	SSI	Algorithm	SSI
Lee Filter	0.9656	Bayes-Proposed	0.9315
Kuan Filter	0.9618	Mod-BayesShrinkage-Soft	0.9198
Frost Filter	0.9672	Mod-BayesShrink-Proposed	0.9123
Guided Filter	0.9345	NormalShrinkage-Soft	0.9196
Bilateral Filter	0.9420	NormalShrinkage-Proposed	0.9167
SRAD	0.9447	Neigh-Shrink-Sure	0.9170
VisuShrinkage-soft	0.9310	Hybrid-Zhang	0.9166
VisuShrinkage-Proposed	0.9283	Hybrid-Proposed	0.9109
BayesShrinkage-Soft	0.9304		

Table 9.12 Comparison of time taken by different de-noising techniques for real ultrasound images (Test Image-4)

Algorithm	Test Image 1	Test Image 2	Test Image 3	Test Image 4	Test Image 5
Lee Filter	0.3685	0.3808	0.4752	0.4588	0.4396
Kuan Filter	0.4480	0.4626	0.5189	0.501	0.4629
Frost Filter	0.8764	0.9157	1.159	1.1191	0.9852
Guided Filter	0.4486	0.4651	0.6372	0.6157	0.5007
Bilateral Filter	0.3573	0.3692	0.4132	0.3738	0.3448
SRAD Filter	0.6237	0.6445	0.8903	0.8546	0.7375
Neigh-Shrink-Sure	0.5739	0.593	0.8756	0.843	0.7261
Hybrid-Zhang	0.8968	0.9276	1.4327	1.3819	1.2714
Hybrid-Proposed	0.9320	0.9631	1.5183	1.462	1.399

References

Adamo F, Andria G, Attivissimo F, Lanzolla AML, Spadavecchia M (2013) A comparative study on mother wavelet selection in ultrasound image denoising. Measurement 46(8):2447–2456

Ando K, Nagaoka R, Hasegawa H (2020) Speckle reduction of medical ultrasound images using deep learning with fully convolutional network. Japanese J Appl Phys 59(SK), SKKE06

Balocco S, Gatta C, Pujol O, Mauri J, Radeva P (2010) SRBF: Speckle reducing bilateral filtering. Ultrasound Med Biol 36(8):1353–1363. https://doi.org/10.1016/j.ultrasmedbio.2010.05.007

Carovac A, Smajlovic F, Junuzovic D (2011) Application of ultrasound in medicine. Acta Informatica Med 19(3):168

Chan V, Perlas A (2010) Basics of ultrasound imaging. In: Atlas of ultrasound-guided procedures in interventional pain management. Springer, New York, pp 13–19

Chang SG, Bin Yu, Vetterli M (2000) Adaptive wavelet thresholding for image denoising and compression. IEEE Trans Image Process 9(9):1532–1546

Chen GY, Bui TD, Krzyzak A (2005) Image denoising using neighbouring wavelet coefficients. In: 2004 IEEE international conference on acoustics, speech, and signal processing. IEEE. https://doi.org/10.1109/icassp.2004.1326408

Chen Y, Zhang M, Yan H-M, Li Y-J, Yang K-F (2019) A new ultrasound speckle reduction algorithm based on superpixel segmentation and detail compensation. Appl Sci 9(8):1693. https://doi.org/10.3390/app9081693

Choi H, Jeong J (2020) Despeckling algorithm for removing speckle noise from ultrasound images. Symmetry 12(6):938

Czerwinski RN, Jones DL, O'Brien WD (1995) Ultrasound speckle reduction by directional median filtering. In: Proceedings of international conference on image processing, IEEE Comput Soc Press

Donoho DL, Johnstone IM (1994) Ideal spatial adaptation by wavelet shrinkage. Biometrika 81(3):425–455

Donoho DL, Johnstone IM (1995) Adapting to unknown smoothness via wavelet shrinkage. J Am Statistical Assoc 90(432):1200–1224. https://doi.org/10.1080/01621459.1995.10476626

Elyasi I, Zarmehi S (2009) Elimination noise by adaptive wavelet threshold. World Acad Sci 32(1):462–466

Frost VS, Stiles JA, Shanmugan KS, Holtzman JC (1982) A model for radar images and its application to adaptive digital filtering of multiplicative noise. IEEE Trans Pattern Anal Mach Intelligence PAMI-4(2):157–166

Gonzalez RC, Woods RE (2008) Digital image processing. Prentice Hall, Upper Saddle River, NJ

Guo Y, Wang Y, Hou T (2011) Speckle filtering of ultrasonic images using a modified non local-based algorithm. Biomed Signal Process Control 6(2):129–138

He K, Sun J, Tang X (2013) Guided image filtering. IEEE Trans Pattern Anal Mach Intelligence 35(6):1397–1409. https://doi.org/10.1109/TPAMI.2012.213

Jain L, Singh P (2020) A novel wavelet thresholding rule for speckle reduction from ultrasound images. J King Saud Univ—Comput Information Sci. https://doi.org/10.1016/j.jksuci.2020.10.009

Jensen JA (1996) Field: a program for simulating ultrasound systems. In: 10th Nordic-Baltic conference on biomedical imaging published in medical and biological engineering and computing, vol 34, issue 1, pp 351–353

Jensen JA, Svendsen NB (1992) Calculation of pressure fields from arbitrarily shaped, apodized, and excited ultrasound transducers. IEEE Trans Ultrasonics Ferroelectrics Frequency Control 39(2):262–267

Joseph WG (1976) Some fundamental properties of speckle. J Optical Soc Am 66(11):1145–1150

Kaur L, Gupta S, Chauhan R (2002) Image denoising using wavelet thresholding. In: International conference on computer vision, graphics and image processing

Kuan DT, Sawchuk AA, Strand TC, Chavel P (1985) Adaptive noise smoothing filter for images with signal-dependent noise. IEEE Trans Pattern Anal Mach Intelligence PAMI-7(2):165–177

Lee J-S (1980) Digital image enhancement and noise filtering by use of local statistics. IEEE Trans Pattern Anal Mach Intelligence PAMI-2(2):165–168

Lin Z, Zhang L, Mou X, Zhang D (2011) FSIM A feature similarity index for image quality assessment. IEEE Trans Image Process 20(8):2378–2386

Loupas T, McDicken WN, Allan PL (1989) An adaptive weighted median filter for speckle suppression in medical ultrasonic images. IEEE Trans Circuits Syst 36(1):129–135

Perona P, Malik J (1990) Scale-space and edge detection using anisotropic diffusion. IEEE Trans Pattern Anal Mach Intelligence 12(7):629–639

Qiu F, Berglund J, Jensen JR, Thakkar P, Ren D (2004) Speckle noise reduction in SAR imagery using a local adaptive median filter. GIScience Remote Sens 41(3): 244–266

Ratliff ST (2013) Webb's physics of medical imaging. Med Phys 40(9) (2nd edn)

Real ultrasound images database collected from Geldrsevallei hospital, the Netherland. https://ultrasoundcases.com/Case-List.aspx?cat=229. Accessed on 21–02–2016

Sagheer MSV, George SN (2017) Ultrasound image despeckling using low rank matrix approximation approach. Biomed Signal Process Control 38:236–249. https://doi.org/10.1016/j.bspc.2017.06.011

Sang Y-F, Wang D, Wu J-C, Zhu Q-P, Wang L (2009) Entropy-based wavelet de-noising method for time series analysis. Entropy 11(4):1123–1147

Sattar F, Floreby L, Salomonsson G, Lovstrom B (1997) Image enhancement based on a nonlinear multiscale method. IEEE Trans Image Process 6(6):888–895

Sheng Y, Xia Z-G (2004) A comprehensive evaluation of filters for radar speckle suppression. In: IGARSS '96. 1996 international geoscience and remote sensing symposium. IEEE

Wang Z, Bovik AC, Sheikh HR, Simoncelli EP (2004) Image quality assessment: from error visibility to structural similarity. IEEE Trans Image Process 13(4):600–612

Wells PNT, Halliwell M (1981) Speckle in ultrasonic imaging. Ultrasonics 19(5):225–229

Wen H, Qi W (2015) Enhancement and denoising method of medical ultrasound image based on wavelet analysis and fuzzy theory. In: 2015 Seventh international conference on measuring technology and mechatronics automation IEEE

Yongjian Yu, Acton ST (2002) Speckle reducing anisotropic diffusion. IEEE Trans Image Process 11(11):1260–1270. https://doi.org/10.1109/tip.2002.804276

Zhang J, Lin G, Wu L, Wang C, Cheng Y (2015) Wavelet and fast bilateral filter based de-speckling method for medical ultrasound images. Biomed Signal Process Control 18:1–10. https://doi.org/10.1016/j.bspc.2014.11.010

Zhang J, Lin G, Wu L, Cheng Y (2016) Speckle filtering of medical ultrasonic images using wavelet and guided filter. Ultrasonics 65:177–193. https://doi.org/10.1016/j.ultras.2015.10.005

Zhou W, Bovik AC (2009) Mean squared error: love it or leave it? A new look at signal fidelity measures. IEEE Signal Process Mag 26(1):98–117

Chapter 10
Poisson Noise-Adapted Total Variation-Based Filter for Restoration and Enhancement of Mammogram Images

Sneha Tiruwa, R. B. Yadav, and Nikhil Singh

Abstract In this manuscript, we introduce a total variation to eradicate noise from an image which the Poisson noise has corrupted. This method regulates the total variation in order to preserve edges. Apart from that, it employs the fidelity of the data term that is better suited to the Poisson noise. A diversity of approaches for denoising mammogram pictures has been described, each with its own assumptions, advantages, and limitations. The efficacy of the filters has also been compared using criteria such as MSE and PSNR. Like the Poisson noise, the result of this regularization is also signaling dependent. The design steps include preprocessing. The preprocessing techniques include manual cropping of original mammograms to remove background details, quantum noise reduction, and contrast enhancements. A modified TV-based filter with the Poisson noise adaptation is utilized to reduce quantum noise.

Keywords Mammogram · Modified TV filter · Poisson noise · TV regularization

10.1 Introduction

Image denoising is a matter of extreme importance when it comes to mathematical image processing. The basic idea lies in converting a noisy image 'f' into a noiseless image 'i', which is the solution of the comparable inverse problem. Out of many algorithms used for reconstruction of 'X' from 'f', the one that of Rubin, Osher, and

S. Tiruwa
Department of Electronics and Communication, G.B. Pant Institute of Engineering and Technology, Pauri Garhwal, India

R. B. Yadav (✉)
Department of Electronics and Communication, G. B. Pant Institute of Engineering and Technology (Formerly G. B. Pant Engineering College), Pauri Garhwal, India

N. Singh
Department of Electronics and Communication Engineering, Delhi Technological University (Formerly Delhi College of Engineering), New Delhi, India
e-mail: nikhil_2k19phdec04@dtu.ac.in

© The Author(s), under exclusive license to Springer Nature Singapore Pte Ltd. 2022
N. Kumar et al. (eds.), *Advance Concepts of Image Processing and Pattern Recognition*,
Transactions on Computer Systems and Networks,
https://doi.org/10.1007/978-981-16-9324-3_10

Fatima (Rudin et al. 1992) is the most successful. Total variation regularization is employed by this algorithm (Srivastava et al. 2012; Kwak et al. 2018). The following is the ROF model solution:

$$F(I) := \int_{\Omega} |\Delta I| + \lambda/2 \int_{\Omega} |f - I| \tag{10.1}$$

In the above equation, the image domain is Ω, and the parameter to be selected is λ. Initially, the equation represents the regularization term, and the primary data is the fidelity term. Or in signal independent, additive Gaussian noise (Green 2002), the ROF model is best suited. Noise in many of the important data is signal dependent and follows the Poisson distribution, for example, in radiography. Mainly, we determine signals in radiography by using the photon counting method. This emphasizes the signal's quantized and non-Gaussian characteristics, making it extremely difficult to eliminate the noise of this nature (Tiruwa and Yadav 2018).

As a result, we use a variety of strategies to accomplish this, but they all rely on the assumption that a small number of wavelet expansion coefficients can accurately define the underlying intensity function. In (Kervrann and Trubuil 2004), authors employed an adaptive windowing approach.

We offer a variational TV regularized (Lee and Lee 2019; Lee et al. 2019) denoising model along the lines of ROF in this model, but we modify it to work with the Poisson noise. We encounter the effects of having a dimensionally varying regularization parameter in the further part of the project.

In the context of Mumford–Shah regularization (Vanzella et al. 2004), the author used a self-adapting parameter technique. Also, Wong and guan (1998) used a neural network approach after the purpose of linear image filtering. Once they assumed noise variance was locally constant, the authors in Reeves (1994) approximated the local parameter values from each iterate they used in the restoration procedure using Laplacian filtering. Authors in Wu et al. (2004) estimated both the parameter values along with linear regularization operation. The previously described approaches required separate computations to approximate parameters, as we discovered (Le et al. 2007). On the other hand, the spatially varying parameter is a result of applying the data fidelity term in our scenario. It was also consistent with the probabilistic noise model.

We make some assumptions for our convenience, such as regarding 'F' as integer values used. From the Poisson distribution with mean and standard deviation SD:

$$X_{\text{SD}}(r) = \frac{p^{-\text{SD}}\mu^r}{r!} \quad r \geq 0. \tag{10.2}$$

Bayes' law is implied to identify the image 'X' (Green 2002), which gives the detailed observed image, according to which:

$$D(I|f) = \frac{\text{d}(f|I)\text{d}(I)}{\text{d}(f)} \tag{10.3}$$

$d(f/I)d(I)$ Thus, to maximize, presuming the Poisson noise, forever $j \in \mu$ we have

$$d(f(x)|X') = d_{I(x)}(f(x)) = \frac{e^{-I(x)} X'(x)^{f(x)}}{f(x)!} \tag{10.4}$$

Presume that f values at the pixels $\{x_i\}$ are independent and that region μ has been pixelated (Bonettini and Ruggiero 2012; Bonettini et al. 2017). Then

$$d(f|X') = \prod_i \frac{e^{-X'(x_i)} X'(x)^{f(x_i)}}{f(x_i)!} \tag{10.5}$$

Considering prior distribution, TV regularization can be found:

$$f(X') = \exp\left(-\beta \int_\Omega |\nabla X'|\right) \tag{10.6}$$

where β is the regularization parameter. $d(f/I)d(I)$ Rather than increasing the $-log(d(f/I))(d(I))$ function is minimized, the result we obtain is thus minimizer of

$$\sum_i (X'(p_i) - f(p_i) \log X'(p_i)) + \beta \int_\Omega |\nabla X'| \tag{10.7}$$

It is regarded as a different approximation of the functional equation.

$$Z(X') := \int_\Omega (X' - f \log X') + \beta \int_\Omega |\nabla X'| \tag{10.8}$$

As defined, the functional E on the set of $X' \in QT(\mu)$ such that $\log I \in L(\mu)$; mainly, for X' it is necessary to be positive in all conditions.

For minimizing $Z(X')$, Euler–Langrage equation is

$$0 = \operatorname{div}\left(\frac{\nabla X'}{|\nabla X'|}\right) + \frac{1}{\beta X'}(f - X), \text{ with} \frac{\partial X'}{\partial n} = 0 \text{ on } \partial \Omega \tag{10.9}$$

For minimizing the ROF functional, we equate these functions using the Euler–Langrage equation (Blomgren et al. 1997; Chan et al. 2000),

$$0 = \operatorname{div}\left(\frac{\nabla I}{|\nabla I|}\right) + \lambda(f - I), \text{ with} \frac{\partial I}{\partial \vec{n}} = 0 \text{ on } \partial I \tag{10.10}$$

Equations 10.9 and 10.10 are found to be similar, but with the variable that depends on restored image I. As expected, noise is increasing with image intensity. This local variation is preferred for the Poisson noise. During solving the single uncon-strained minimization problem, the self-adaption occurs automatically. This scheme can further be developed to introduce a 3D imaging model (Kastanis et al. 2008).

10.2 Numerical Result

Gradient descent is preferably used. A strictly straightforward and discretized version of the below-mentioned PDE is implemented as follows:

$$I_t = \text{div}\left(\frac{\nabla I}{|\nabla I|}\right) + \frac{1}{\beta I}(f - I), \text{ with} \frac{\partial I}{\partial \vec{n}} = 0 \text{ on } \partial\Omega \qquad (10.11)$$

Computation of derivatives is done with standard centered difference approxima-tion.

For small, positive E, the numerical quantity (ΔI) is used instead of $\sqrt{|\Delta I|2 + \varepsilon}$. Constant time steps are taken for time evolution till there is a sufficiently small change in i. To compare our proposed model, implementation of a similar procedure is used for the ROF model.

An example of Fig. 10.1 inside the square frame of intensity ten circles is enclosed with intensities of 70,135, and 200, with background intensity of 5. The Poisson noise is depending on absolute image intensities, yet no parameter is linked with the Poisson noise. With the increment in the intensity of the image, the proportion of noise and region of the image increases. Parameter λ, which lies on the result, is chosen regarding the discrepancy principle (Engl et al. 1996; Zanella et al. 2009). According to this rule, the mean square difference between noisy data and the reconstructed image should equal the noise variance. In other words, images that are chosen should be most consistent among the entire possibly reconstructed image.

The noise variance would have to be approximated if the original image was not accessible. In the given condition, the resulting value of γ, i.e., 0.04, provides the image. The frame can be preserved by boosting gamma by 0.04, as the frame is totally blown out and changes from the backdrop by 7.33, i.e., the standard deviation of noise (Hirakawa 2007). As a result, the noise in the higher intensity zones is maintained and not washed off.

For our Poisson modified TV model, parameter β is chosen according to the discrepancy rule: The value of $\int X' - f \log X'$ should match the original image (Osher et al. 2005). This resulted in β with a value of 0.25 with signal-dependant $\lambda = 1/\beta$. Hence, we have an effective λ of 0.08 for the background, 0.4 for the frame, and 0.02 for the center. Value 0.02 gives stronger regularization than that of ROF from the discrepancy principle (Wen and Chan 2011; Rose et al. 2015). Moreover, in images, qualitative line outs of the images could be clearly seen.

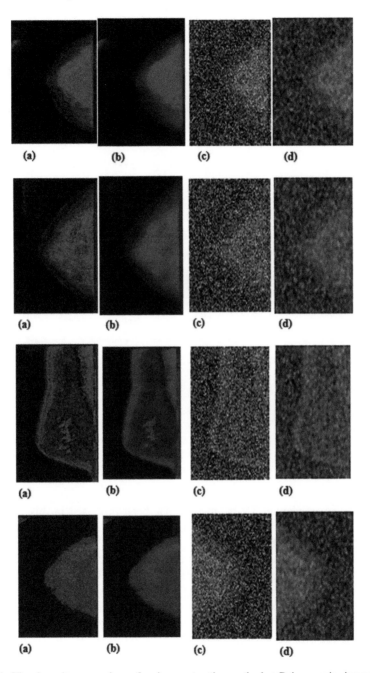

Fig. 10.1 Visual results comparison of various restoration methods **a** Poisson noise image, **b** total variation, **c** median filter, **d** weiner filter

Fig. 10.2 MSE-based comparison of filters

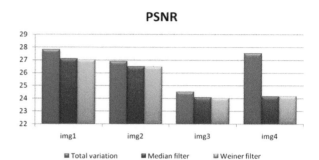

Fig. 10.3 PSNR-based comparison of filters

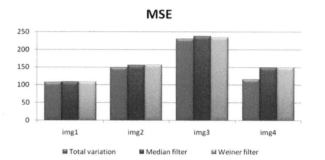

10.3 Results

See Figs. 10.2 and 10.3.

10.4 Conclusion

The Poisson noise-adapted TV-based filter has been proposed in this manuscript that focuses on the restoration of noisy mammogram images. The performance of different image restoration approaches is measured with the help of PSNR and MSE, and the Poisson noise was introduced artificially in the test images. We added the Poisson noise to the images in order to calculate the MSE and PSNR of the observation model. MATLAB was used to implement all of the techniques. Under this scheme, the regularization parameter is considered suitably. We also observed that the resulting regularization allows features to be saved in the low-intensity regions of the images. It can be concluded from the achieved result that the Poisson noise-adapted TV regularization-based method outperforms other image restoration methods. This approach is capable of restoring good image quality while maintaining strong edge preservations.

References

Besbeas P, Fies ID, Sapatinas T (2004) A comparative simulation study of wavelet shrinkage estimators for Poisson counts. Int Stat Rev 72:209–237

Blomgren P, Chan TF, Mulet P, Wong CK (1997, October) Total variation image restoration: numerical methods and extensions. In: Proceedings of international conference on image processing, vol 3. IEEE, pp 384–387

Bonettini S, Ruggiero V (2012) On the convergence of primal–dual hybrid gradient algorithms for total variation image restoration. J Math Imaging Vis 44(3):236–253

Bonettini S, Loris I, Porta F, Prato M, Rebegoldi S (2017) On the convergence of a linesearch based proximal-gradient method for nonconvex optimization. Inverse Prob 33(5):055005

Chan T, Marquina A, Mulet P (2000) High-order total variation-based image restoration. SIAM J Sci Comput 22(2):503–516

Donoho D (1993) Nonlinear wavelet methods for recovery of signals, densities and spectra from indirect and noisy data. In: Proceedings of symposia in applied mathematics: different perspectives on wavelets. American Mathematical Society, pp 173–205

Engl HW, Hanke M, Neubauer A (1996) Regularization of inverse problems, vol 375. Springer Science & Business Media

Evans LC, Gariepy RF (1992) Measure theory and fine properties of functions. CRC Press, Boca Raton

Green M (2002) Statistics of images, the TV algorithm of Rudin-Osher-Fatemi for image denoising and an improved denoising algorithm. CAM Report 02-55, UCLA, October 2002

Hirakawa K (2007, September) Signal-dependent noise characterization in Haar filterbank representation. In: Wavelets XII, vol 6701. International Society for Optics and Photonics, p 67011I

Jonson E, Huang C-S, Chan T (1998) Total variation regularization in positron emission tomography. CAM Report 98-48, UCLA, November 1998

Kastanis I, Arridge S, Stewart A, Gunn S, Ullberg C, Francke T (2008, July) 3D digital breast tomosynthesis using total variation regularization. In: International workshop on digital mammography. Springer, Berlin, Heidelberg, pp 621–627

Kervrann C, Trubuil A (2004) An adaptive window approach for poisson noise reduction and structure preserving in confocal microscopy. In: International symposium on biomedical imaging (ISBI'04), Arlington, VA, April 2004

Kolaczyk E (1999) Wavelet shrinkage estimation of certain poisson intensity signals using corrected thresholds. Statist Sinica 9:119–135

Kwak HJ, Lee SJ, Lee Y, Lee DH (2018) Quantitative study of total variation (TV) noise reduction algorithm with chest X-ray imaging. J Instrum 13(01):T01006

Le T, Chartrand R, Asaki TJ (2007) A variational approach to reconstructing images corrupted by Poisson noise. J Math Imaging Vis 27(3):257–263

Lee S, Lee Y (2019) Performance evaluation of total variation (TV) denoising technique for dual-energy contrast-enhanced digital mammography (CEDM) with photon counting detector (PCD): Monte Carlo simulation study. Radiat Phys Chem 156:94–100

Lee S, Park SJ, Jeon JM, Lee MH, Ryu DY, Lee E, Lee Y et al (2019) Noise removal in medical mammography images using fast non-local means denoising algorithm for early breast cancer detection: a phantom study. Optik 180:569–575

Osher S, Burger M, Goldfarb D, Xu J, Yin W (2005) An iterative regularization method for total variation-based image restoration. Multiscale Model Simul 4(2):460–489

Reeves S (1994) Optimal space-varying regularization in iterative image restoration. IEEE Trans Image Process 3:319–324

Rose S, Andersen MS, Sidky EY, Pan X (2015) Noise properties of CT images reconstructed by use of constrained total-variation, data-discrepancy minimization. Med Phys 42(5):2690–2698

Rudin L, Osher S, Fatemi E (1992) Nonlinear total variation based noise removal algorithms. Physica D 60:259–268

Rudin LI, Osher S (1994) Total variation based image restoration with free local constraints. ICIP 1:31–35

Srivastava S, Sharma N, Srivastava R, Singh SK (2012, December) Restoration of digital mammographic images corrupted with quantum noise using an adaptive total variation (TV) based nonlinear filter. In: 2012 international conference on communications, devices and intelligent systems (CODIS). IEEE, pp 125–128

Timmermann K, Novak R (1999) Multiscalemodeling and estimation of Poisson processes with applications to photon limited imaging. IEEE Trans Inf Theor 45:846–852

Tiruwa S, Yadav RB (2018, October) Comparing various filtering techniques for reducing noise in MRI. In: 2018 international conference on sustainable energy, electronics, and computing systems (SEEMS). IEEE, pp 1–5

Vanzella W, Pellegrino FA, Torre V (2004) Self adaptive regularization. IEEE Trans Pattern Anal Mach Intell 26:804–809

Wen YW, Chan RH (2011) Parameter selection for total-variation-based image restoration using discrepancy principle. IEEE Trans Image Process 21(4):1770–1781

Wong H, Guan L (1998) Adaptive regularization in image restoration by unsupervised learning. J Electron Imaging 7:211–221

Wu X, Wang R, Wang C (2004) Regularized image restoration based on adaptively selecting parameter and operator. In: 17th international conference on pattern recognition (ICPR'04), Cambridge, UK, August 2004, pp 602–605

Zanella R, Boccacci P, Zanni L, Bertero M (2009) Efficient gradient projection methods for edge-preserving removal of Poisson noise. Inverse Prob 25(4):045010

Chapter 11
Implementation of Mathematical Morphology Technique in Binary and Grayscale Image

Arun Kumar, Sumit Chakravarty, Manoj Gupta, Imran Baig, and Mahmoud A. Albreem

Abstract Morphological picture supervision is a collection of non-direct responsibilities acknowledged with the shape or morphology of high points in an image. Morphological responsibilities depend just on the inclusive in treating of pixel respects, not on their arithmetical qualities, and in this way are predominantly fit to the handling of corresponding images. Morphological tasks can equally be functional to grayscale images with the culmination goal that their light exchange dimensions are ambiguous. Subsequently, their absolute pixel reverences are of no or minor scheming. The proposed work focused 3 on enhancing the image by utilizing the morphological technique: erosion, dilation, opening and closing. The arrangement of elements determined the increase of binary pixel in an image. When erosion is applied to the grayscale image, it enhanced the intensity of an image through enchanting the region extreme after fleeting the constructing part over the object. Outcomes of the technique carried out to several sides of an object indicate that abstract data of the reference object can be effortlessly obtained. The proposed technique can be utilized to develop an appropriate islanding method and can be used to recognize the weak associations in the object.

Keywords Erosion · Dilation · Opening · Closing · Image processing

A. Kumar (✉) · M. Gupta
Department of Electronics and Communication, JECRC University, Jaipur 303905, India
e-mail: arun.kumar@jecrcu.edu.in

S. Chakravarty
Department of Electrical and Computer Engineering, Kennesaw State University, Kennesaw, GA, USA
e-mail: schakra2@kennesaw.edu

I. Baig
Department of Electrical and Computer Engineering, College of Engineering, Dhofar University, Salalah 211, Sultanate of Oman

M. A. Albreem
Department of Electrical Engineering, College of Engineering, University of Sharjah, Sharjah, UAE
e-mail: mahmoud.albreem@asu.edu.com

N. Kumar et al. (eds.), *Advance Concepts of Image Processing and Pattern Recognition*,
Transactions on Computer Systems and Networks,
https://doi.org/10.1007/978-981-16-9324-3_11

11.1 Introduction

Morphology is the process to define dimension, aspect and the alignment of the image. It is the method to amputate the information lesser than the certain mentioned threshold. This technique is mostly suited to measure the sets of the two-dimensional mark of an image. It has found its application to abstract, filter and de-noise signal (Clienti et al. 2008; Torres-Huitzil 2013). Binarization of image involves global threshold techniques to represent the grayscale image to black–white image. The pixel scale of the grayscale image is 0–255. However, it is a difficult task to set the universal threshold value for all images. Morphological techniques are carried out to fix the binary picture at the midpoint of every pixel. In order to achieve an efficient result, alternates operation is carried out on the image. The method of color in the certain portion of the image is known as region filling. It is carried to determine the level of the binary image (Tcheslavski 2010; Heijmans 1994). In this work, morphological scheme is utilized to improve the surrounding of the image. The proposed work is implemented by using MATLAB software (Radhakrishnan et al. 2005). The authors proposed a technique based on the characteristics of the image. The simulation results reveal that the proposed scheme efficiently utilized the properties of an image as compared to the conventional techniques (Sundararajan 2017; Kaushal 2017; Naganandhini and Shanmugavadivu 2019). The authors introduced a new method to enhance the background of an image. The proposed work was accomplished by utilizing a rated capacity line scheme for creation of an image (Vincent 1992; Deforges et al. 2013). The proposed work deals with the comprehensive study of the morphology techniques, for enhancing the features of an image. Further, method to represent a binary image to grayscale image is also carried out (Dokadal and Dokladalova 2011; Gibson et al. 2013a; Gonzales and Woods 2002). In this work, a nonlinear filtration is applied to enhance the dimension of an image. The outcomes of the proposed technique reveal that the performance of the proposed technique is better than the conventional image processing techniques (Avery et al. 2007; Serra 1994; U.Scot.E 1998). The authors introduced a novel technique to analyze the brain images. The proposed work is executed in two steps. In the first step, the location of the brain is identified, and in the second step, morphological techniques are carried out (Fang et al. 2012; Qin et al. 2019; Gong et al. 2019). In this work, we focus on closing, opening, dilation and erosion of an image. The main objective of the proposed work is to enhance the image by utilizing the morphological technique.

11.2 System Model

In this work, we have discussed the morphological technique such as dilation, erosion, closing and opening.

11.2.1 Dilation

It is a method of enhancing the pixel properties of the binary images. The arrangement of elements determined the increase of binary pixel in an image (Li et al. 2019; Zhao et al. 2012). The dimension of an element used for expanding the images must be smaller than a reference pixel. Mathematically, it is represented as follows (Haralick and Zhuang 1987; Bartovsky et al. 2014; Gil and Kimmel 2002):

$$Y \oplus A = \sum x \left| (\widehat{A})_x \cap Y \right| \subseteq Y \tag{11.1}$$

$\widehat{A}$ is the image obtained after the rotation of image A, and Y is the reference image.

11.2.2 Erosion

It is one of the techniques used in morphology for enhancement of binary and grayscale image. Let us consider a binary image represented by I (Clienti et al. 2008; Gil and Kimmel 2000; Dokladal and Dokladalova 2011). The erosion of I by element K ($I \theta K$) generates an advanced image represented as $h = I \theta K$, in the coordinates of (x, y).

11.2.3 Opening and Closing

The opening is one of the important operations of morphology. It is carried out to clip the binary image pixels of the coordinate (Torres-Huitzil 2013; Deforges et al. 2013; Gibson et al. 2013b; Kumar et al. 2010). It is given as follows:

$$A \circ B = (A \theta B) \oplus B,$$

where θ represents erosion and $\oplus$ dilation.

The closing of an image is carried out with the opening. The main objective of the process is to remove the unwanted noise in the binary image (Patidar et al. 2010; Gupta et al. 2011, 2018a, 2018b, 2020). Mathematically, it is given as follows:

$$A \bullet B = (A \oplus B) \theta B$$

11.3   Simulation Results

The main objective of the proposed work is to enhance the image by utilizing the morphological technique. Dilation is carried away to enhance the dimension of an image, satisfying in holes and damaged zones, and relating parts that are disconnected by spaces lesser than the proportion of the constructing part. Figure 11.1a indicates the original binary image. A morphological dilation operation is performed on the original image. Figure 11.1b represents the dilated binary image. In Fig. 11.1a, the area represented by white color was originally a black pixel. Hence, it is concluded that the dilation operation enlarges the white pixel of a binary image. To further analyze the dilation operation, it is performed on grayscale image, as shown in Fig. 11.2. It is noted that when erosion is applied to the binary image, it attaches the elements that are detached by spaces lesser than the constructing component and improves pixels to the boundary of all image. When erosion is applied to the grayscale image, it enhanced the intensity of an image through enchanting the region extreme after fleeting the constructing part over the object.

Fig. 11.1 **a** Original image, **b** binary image

(a)

(b)

Fig. 11.2 Original and
dilated grayscale image

Erosion is carried out to decrease the dimension of an image and separates minute irregularities by detracting images with an area lesser than the constructing component. Figures 11.3 and 11.4 represent the eroded version of dilated binary and grayscale image. The erosion decreases the intensity and dimension of a grayscale image on a shady surface by enchanting the region lowest when assisting the constructing component over the object. It is noted that when erosion is applied to the binary dilated image, it completely eliminates the objects, lesser than the building component and eradicates the edge pixels from bigger image.

Figure 5a, b represents the morphological opening operation of binary and grayscale images. Opening by disks conveys around a medium that is stable from within; that is, it adjusts corners reaching out away from plain view. The impact is very unique with a square organizing component. As opposed to seeing the opened

Fig. 11.3 Eroded binary
image

Fig. 11.4 Eroded grayscale
image

object itself as the last yield of the preparing, we can take an alternate view. We can consider the set theoretic subtraction of the opening from the reference object.

Figure 6a, b represents the morphological closing operation of binary and grayscale images. The function of closing is opposite as compared to the opening. The variance is filling of minor impositions. It is noted that that the closing operation eliminates the trivial dumps.

11.4 Conclusion

Morphological examination is appropriate for surfaces since as a nonlinear shape-based picture separating system, it exceeds expectations at the misuse of spatial connections among pixels and has countless instruments fit for removing size and shape data. This turns out to be particularly significant while focusing on surface natives. In addition, multiscale morphological devices can be actualized moderately effectively, in this way dealing with surface crude size varieties, while associated morphological administrators speak to a further incredible arrangement of devices fit for abusing pixel networks. Likewise, rather than various insights and Fourier changes, which portray just a surface procedure up to second-arrange qualities, morphological strategies can catch higher-request properties of the spatial irregular procedures. Lastly, if there should be an occurrence of expanded proficiency necessities, morphological devices can be legitimately communicated with Boolean variable-based math and along these lines can be executed on devoted equipment. To put it plainly, given the spatial idea of the existed conventional surface attributes except for generally speaking shading and shading immaculateness, morphology has

Fig. 11.5 Opening
operation: **a** binary, **b**
grayscale

a

b

a significant hypothetical bit of leeway over its direct partners in catching them. In this work, we proposed a morphology operation grounded on the structures of binary and grayscale objects. In the first step, the quality of image in enhanced and mathematical morphology processes is carried out to process the images. Outcomes of the technique carried out to several sides of an object indicate that abstract data of the reference object can be effortlessly obtained. The proposed technique can be utilized to develop an appropriate islanding method and can be used to recognize the weak associations in the object.

Fig. 11.6 Closing morphology: **a** binary image, **b** grayscale

a

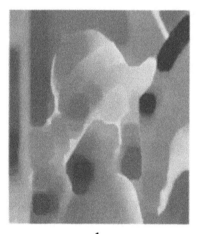

b

References

Avery RP, Zhang G, Wang Y (2007) An investigation into shadow removal from traffic images. University of Washington

Clienti C, Beucher S, Bilodeau M (2008) A system on chip dedicated to pipeline neighborhood processing for mathematical morphology. In: 16th European IEEE conference in signal processing, pp 1–5

Dokadal P, Dokladalova E (2011) Computationally efficient, one-pass algorithm for morphological filters. J Vis Commun Image Represent 22(5):411–420

Fang Z, Yulei M, Junpeng Z (2012) Medical image processing based on mathematical morphology. In: The 2nd international conference on computer application and system modeling

Gibson RM, Ahmadinia A, McMeekin SG, Strang NC, Morison G (2013a) A reconfigurable real-time morphological system for augmented vision. EURASIP J Adv Sig Process 2013(1):1–13

Gibson RM, Ahmadinia A, McMeekin SG, Strang NC, Morison G (2013b) A reconfigurable real-time morphological system for augmented vision. EURASIP J Adv Sig Process 2013(1):1–13

Gil J, Kimmel R (2002) Efficient dilation, erosion, opening, and closing algorithms. IEEE Trans Pattern Anal Mach Intell 24(12):1606–1617

Gil J, Kimmel R (2000) Efficient dilation, erosion, opening and closing algorithms in mathematical morphology and its applications to image and signal processing. In: Goutsias VJ, Vincent L, Bloomberg D (eds). Palo-Alto, USA, June 2000. Kluwer Academic Publisher, pp 301–310

Gong K et al (2019) Iterative PET image reconstruction using convolutional neural network representation. IEEE Trans Med Imaging 38(3):675–685. https://doi.org/10.1109/TMI.2018.286 9871

Gonzales RC, Woods RE (2002) Digital image processing, 2nd ed. Prentice Hall

Gupta M, Upadhyay S, Nagawat AK (2011) Camaera calibration techniques using Tsai's algorithms. Int J Enterprise Comput Bus Syst 1(2):1–9

Gupta M, Taneja H, Chand L (2018a) Performance enhancement and analysis of filters in ultrasound image denoising. Procedia Comput Sci 132:643–652

Gupta M, Taneja H, Chand L, Goyal V (2018b) Enhancement and analysis in MRI image denoising for different filtering techniques. J Stat Manag Syst 21(4):561–568

Gupta M, Lechner J, Agarwal B (2020) Performance analysis of Kalman filter in computed tomography thorax for image denoising. Rec Adv Comput Sci Commun 13(6):1199–1212

Haralick SS, Zhuang X (1987) Image analysis using mathematical morphology. IEEE Trans Pattern Anal Mach Intell 9(4):532–550

Heijmans H (1994) Morphological image operators. In: Advances in electronics and electron physics. Academic Press

Bartovsky J, Dokládal P, Dokladalova E, Georgiev V (2014) Parallel implementation of sequential morphological filters. J Real-Time Image Process 9(2):315–327

Kaushal P (2017) A study of morphological operators on digital colour image. Int J Comput Appl Math 12(2):569–578

Kumar S, Kumar P, Gupta M, Nagawat AK (2010) Performance comparison of median and wiener filter in image de-noising. Int J Comput Appl 12(4):27–31

Li L, Wang LG, Teixeira FL (2019) Performance analysis and dynamic evolution of deep convolutional neural network for electromagnetic inverse scattering. IEEE Antennas Wirel Propag Lett 18(11):2259–2263. https://doi.org/10.1109/LAWP.2019.2927543

Naganandhini S, Shanmugavadivu P (2019) Thresholded and morphological operations based brain image segmentation. Int J Innov Technol Explor Eng (IJITEE) 8(6S3):96–100. https://doi.org/F10170486S319/19©BEIESP

Deforges O, Normand N, Babel M (2013) Fast recursive grayscale morphology operators: from the algorithm to the pipeline architecture. J Real-Time Image Process 8(2):143–152

Déforges O, Normand N, Babel M (2013) Fast recursive grayscale morphology operators: from the algorithm to the pipeline architecture. J Real-Time Image Process 8(2):143–152

Dokládal P, Dokladalova E (2011) Computationally efficient, one-pass algorithm for morphological filters. J Vis Commun Image Represent 22(5):411–420

Patidar P, Gupta M, Shrivastava S, Nagawat AK (2010) Image denoising by various filters for different noise. Int J Comput Appl 9(4):45–50

Qin C, Schlemper J, Caballero J, Price AN, Hajnal JV, Rueckert D (2019) Convolutional recurrent neural networks for dynamic MR image reconstruction. IEEE Trans Med Imaging 38(1):280–290. https://doi.org/10.1109/TMI.2018.2863670

Radhakrishnan P, Sagar BSD, Venkatesh B (2005) Morphological image analysis of transmission systems. IEEE Trans Power Deliv 20(1):219–223. https://doi.org/10.1109/TPWRD.2004.839213

Serra J (1994) Mathematical morphology and its applications to image processing. Kluwer Academic Publishers, Boston

Sundararajan D (2017) Morphological image processing. In: Digital image processing. Springer, Singapore. https://doi.org/10.1007/978-981-10-6113-4_8

Tcheslavski GV (2010) Morphological image processing: grayscale morphology. ELEN 4304/5365 DIP, Spring

Torres-Huitzil C (2013) Fast hardware architecture for grey level image morphology with flat structuring elements. IET Image Proc 8(2):112–121

U.Scot.E (1998) Computer vision and image processing. Prentice Hall, NJ. ISBN 0-13-264599-8

Vincent L (1992) Morphological area openings and closings for greyscale images. In: Proceedings of shape in picture '92, NATO workshop, Driebergen, The Netherlands, September 1992. Springer

Zhao F, Zhang J, Ma Y (2012) Medical image processing based on mathematical morphology. In: The 2nd international conference on computer application and system modeling, pp 948–950

Chapter 12
Design of Advanced Security System Using Vein Pattern Recognition and Image Segmentation Techniques

G. Rajakumar and T. Ananth Kumar

Abstract As a biometric trademark, finger vein established-based technology is exceptionally right for persona clear attestation with high security. The vibe of an individual or lady ID shape is mounted on near infrared (NIR) finger vein. The proposed self-flexible illuminance control test introduced an information mannequin of finger vein imaging and activated into photograph guaranteeing about gear. As considered with the assistance of methods for the circulation of pixels profundity of the gained picture, the proposed figuring may also pick to generally trade the illuminance improvement of lighting. Builds up the illuminance of lights beneath which the thicker bit of the finger constitution is brought and decreases the illuminance of lights beneath which the thin upper bit of the finger body is presented. With this change, the total finger physical makeup ought to be lit up fittingly as respected by method for its thickness transport, and along these lines, the overexposure and underexposure are saved as an unprecedented essential course from suitably. A NIR finger vein picture database containing 2040 photographs is made starting at now. Inside the photograph pre-overseeing arrange, Gabor channels are used to redesign got foul finger vein pictures. Inside the paper, the distinct insistence introduction of the proposed structure is outlined utilizing the revelation cost and, subsequently, the edge task. A despairing depiction-based estimation is utilized to figure the affirmation cost and offers information to situate evaluation. The suggestions show the ampleness of the proposed illuminance control comprises of, and in this way. The whole finger vein system is mainly based on individual irrefutable affirmation.

Keywords Biometric · Finger vein · Security · IoT · Image · Segmentation

G. Rajakumar
Department of Electronics and Communication Engineering, Francis Xavier Engineering College, Tirunelveli, Tamilnadu, India

T. Ananth Kumar (✉)
Department of Computer Science and Engineering, IFET College of Engineering, Villupuram, Tamilnadu, India
e-mail: tananthkumar@ifet.ac.in

N. Kumar et al. (eds.), *Advance Concepts of Image Processing and Pattern Recognition*,
Transactions on Computer Systems and Networks,
https://doi.org/10.1007/978-981-16-9324-3_12

12.1 Introduction

The biometric-embedded architecture might be a mix of biomedical instruments and software programmed acclimated with achieving single task inside a given significant measure, over but then again and on and on, with or excepting coordinated human efforts—introduced a course of action of a system that screens and respond to oversee an external area. A condition related to the architecture through sensors, actuators, and various realities yields interfaces. Embedded structures must meet arranging and uncommon hindrances constrained on that with the assistance of the planet. Biometrics suggests the conspicuous evidence of people through their traits or characteristics (Bracken 2020; Vega et al. 2014). Biometrics is utilized in programming designing as a recognizing verification and access control. It is also acclimated to capture people in packs that are underneath perception. Biometric identifiers are the undeniable, quantifiable attributes acclimated with recognizing and portray individuals (Suresh Kumar et al. 2021). Biometric identifiers are consistently requested as physiological versus social qualities established system which might be a mix of medications and programming (Samuel et al. 2021). It is used to achieve a single task within a given period, aside from joint human efforts. RTOS portrays the methodology for the structure works.

The microcontroller might be a significantly organized chip that contains all the portions, including the controller (Narmadha et al. 2020). It pleasantly is likewise distributed by utilizing consolidating more prominent limits into the CPU chip. Biometrics is utilized to get a handle on the properties of people (Xie et al. 2017). Competitive Gabor response (CGR) data packages include a Gabor filter bank with a range of orientation strengths and magnitudes, as well as information about each bank's peak responses. The authors create a histogram by concatenating the CGM and CGO images for a given image and combining it with the HOGC histogram (HCGR) image (Lu et al. 2014). To overcome the issues, we used a spread of sensors in an exceedingly multi-biometric system. Finger vein might be a biometric approval structure that organizes the vascular occurrence in an exceedingly individual's finger to the late obtained data.

12.2 Related Works

Shende and Dandwate introduced convolutional neural network-based multimodal biometric human authentication in 2020 (Shende and Dandwate 2020). Face, palm veins, and fingerprint channel parts, a single-pixel string, a maximum pooling window size of 2×2, and six-channel bits are used for multimodal biometric human authentication. The structure's presentation is evaluated using a database of faces, palm veins, and top gravings organized by actual percentage (Thiruvikraman et al. 2021). Yong et al. investigated the finger vein pattern recognition technology in conjunction with an FPGA (Yong 2020). As with the other type of biometric affirmation development, the cutting-edge case of the finger vein affirmation has enlisted and confirmed the impediment of average speed. As a result, a finger vein spotlight extraction and planning affirmation computation suitable for FPGA is proposed. The apparatus relationship for finger vein affirmation is organized and completed using three modules: finger vein image ensuring the device, photograph supervising module, and planning show module. Finally, critical estimation is acted out and examined, which expedites the enrollment and validation of cellphone veins (Kumar et al. 2021). Madhusudhan et al. published an article on finger vein-based authentication using deep learning techniques in 2020 as security is one of the significant concerns of recent times (Madhusudhan et al. 2019). Biometric-based techniques are considered as powerfully stable and cautious in confirming a private. Hand-based biometric includes at long last wind up being suitably open all through data assortment. The authors dealt with and getting prepared biometric trademark photographs of the vast number of vendors. It is used to take a look into for more noteworthy affiliations. Critical taking in systems acts as the legend from such conditions. At present, prescribe a one-of-a-kind way of thinking for endorsement of the work of finger vein pictures. They utilized an entire convolutional neural system (CNN) to travel learning. The mannequin has been pre-masterminded on various photographs open on the ImageNet database through ResNet-50 arrangements (Revathy et al. 2019). In 2020, El Mehdi Cherrat and Bouzahir conveyed a convolutional neural systems strategy for multimodal biometric ID shape utilizing the combo of a mind-blowing engrave, finger vein, and face pictures (El mehdi Cherrat and Bouzahir 2020). The basic idea behind this paper is to support a crossbreed course of movement of joining the effect of ground-breaking tree models: Convolutional neural structure (CNN), softmax, and random backwoods (RF) classifier issue to multi-biometric finger impression, finger vein, and face-perceiving check framework (Kumar et al. 2020). In 2019, Shaveta Dargon et al. introduced a detailed report on application-shaped biometric confirmation frameworks based on expert physiological and social modalities (Dargan and Kumar 2020). The Authors explored a kind of contraction for biometric innovations based on social finger effect and finger vein plans, which provides well-being and revocability format (Hwang and Park 2018). The multi-biometric form depended on focal points based on modern engraving aspects and picture-based finger vein elements, and along these lines, the portion stage blend inclinations are used to upgrade protection consistency. Their recommendation of enhanced partial-discrete cosine transform (EP-DFT) depended on a non-invertible exchange, giving the buyer an agreeable and vital improvement in protection. Matsuda et al. suggested a finger vein approach biometric interface configuration. They introduced one-of-a-kind picking up knowledge on a technique that is known as DBC (Matsuda et al. 2016). The approach integrated the required relation map to use the grandeur in twofold organizations to manage vein attributes. Using synchronized data and SVM, DBC was seen to be cut off and shorter besides. The PolyU database and MLA database were selected and successfully executed (Sasireka and Rajesh 2014; Abitha et al. 2020). Wang et al. presented a solid and fruitful structure for one in everything about structure engrave attestation on an essential level for asset restricted applications (Xiang et al. 2016). It depended upon the midway Hadamard exchange strategy for the masterminding of cancelable one in everything about structure finger sway plans. This framework depended upon the cancelable one in everything about structure finger influence codecs and stochastic segment remodels that gave revocability, top of the line arrangement, non-invertibility, and execution and affirmed high-caliber as separated and condition of convincing work of art. Qiu et al. suggested another biometric shape finger vein procedure to trigger concern to pseudo indirect transformers and double window constraints (Qiu et al. 2017). Here, planar imaging and double windows contributed to the pixelation of the finger and the finger venous situation. The authors proposed a one-of-a-kind one-in everything about kind engraving-based attestation structure and affirmation invent that relies upon the tempestuous encryption by method for utilizing Murillo-Escobar's be tallied and the contraption legitimately. They utilized a 32-piece microcontroller for secure endorsement frameworks and inserted pro structures, and introduced a hard and fast security evaluation at every client and real level. A severe review of the engraving assertion framework with different applications is done. For instance, mental attacker prominent affirmation, hoodlum evaluation, and other firmly shut parts of the beneficial execution of unique finger impact biometric structure, a first-class good incredible engraving picture are required (Pakutharivu and Srinath 2017). A face-biometric authentication gadget (ASM) model and principal component analysis (PCA) methods were introduced by Yim et al. (2015), who obtained a facial model (Yim et al. 2015). AdaBoost and histogram equalization (HE) techniques were used to enhance the picture's understanding. The radial basis function neural network proved to be adequate proof in the test. By employing differential evolution (DE), all of the conventional parameters for systemic action are also evaluated and reorganized, and promising returns and accuracy rates are transmitted. These parameters include learning rates, fluctuation coefficients, and the office coefficient used by the FCM. The Authors proposed a finger-based biometric system based on a finger vein blend and a method for determining the geometry of the finger (Asaari et al. 2016) that could be used for each of the procedures described above. They made their predictions by examining the association between BLPOC variables that were not deterrent, flexible, or rescaling. The width center contour distance (WCCD) was used in this technique, a geometrical component that combines the finger width and the center contour distance (CCD), allowing for greater precision in confirming the finger geometry. Lin et al. published a classifier for the enthralling gray relativity analysis (GRA) graving assertion in 2011 (Lee and Lin 2011). They were the first to propose using a specific vein on the fingers as a new form of identification, which was proposed by Kono et al. Many studies have considered veins in their analyses (Kono et al. 2002). The acquisition of images is the first step in the process. Unfortunately, high-quality finger vein images alone are insufficient for diagnosing and treating the condition; pretreatment is required as part of the process. System architectures based on deep learning have displaced traditional systems; convolutional neural networks (CNNs) exhibit precision and speed, which are characteristics of conventional algorithms (Lee et al. 2011). The multilayer perceptron (MLP) was used to examine seven individuals and fourteen test samples. The MLP was found to have a 93% accuracy rate in identifying the subjects. In a study conducted by Wang et al. (2012), SVM was applied to a database of ten people, each of whom had 80 images. By using an optical image in each type of finger impression, one in each of them, some finger influence images are taken. To reassure refining influence, image update, double-photograph, and perspective area adjustments, the modern picture preprocessing technique is used to estimate the fractal dimension for the 2D two-figure picture; Katz's figure and Weierstrass cosine function (WCF) have been used to clarify

12.3   Methodology

The vascular model during an individual's finger to beginning late procured data in this proposed work. The introduced architecture could be a blend of medications, and programming will not achieve alone task internal a given period, forever ceaselessly, with or excepting human joint undertakings. The introduced course of movement of a PC system that screens respond to direct a terrace space. A condition related to the system through sensors, actuators, and various measurements yields interfaces. Embedded systems must meet engineering and particular confinements constrained on that utilizing the planet. Biometrics surmises the undeniable take a look at individuals by using their characteristics or characteristics. Biometrics is utilized in programming functioning as a visual affirmation and gets the privilege of passage to direct. It is furthermore will not to see people in packs that are under acumen. Biometric identifiers are the prominent, quantifiable credits that will not name and delineate individuals, referred to as physiological versus social properties. The introduced system could be a combo of gear and programming. It is used to achieve a particular venture internal a given time run, again and again, ceaselessly, with or without human joint undertakings. RTOS portrays the system for the structure works. The microcontroller is an orchestrated chip that contains all the sections comprising the controller. It will be cleaned by utilizing merging more cutoff factors into the CPU chip. Biometrics is utilized to work out the qualities of people. For vanquishing the limits, use an assortment of sensors during a multi-biometric structure. The Finger vein could be a biometric guaranteeing shape that assortments out.

12.4   Proposed System

A real-time finger vein recognition system for authentication on mobile devices is proposed and shown in Fig. 12.1. The system is referenced on a given arranged and issued a one-size-finger vein validation count. The proposed system integrates three interface devices: image procurement interface, integrated mainboard, and human–machine communication module.

The photograph-acquisition module, responsible for acquiring complete images of the finger veins, serves as the system's structural structure and foundation. The vein in the finger may be a promising biometric structure for the precise individual

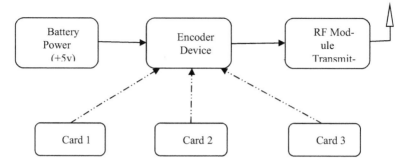

Fig. 12.1 Block diagram of transmitter section

assertion of the certificate's security and convenience, especially when combined with other biometric structures. Typically, the vein is hidden by the physical makeup and is therefore imperceptible to standard eyes. This indicates to the vein that it is trying to make or take something. In addition to ensuring the customer's complete understanding and cleanliness, the unnecessary and contactless finger vein grab is consistently commendable. To create a stay body, it is necessary to use a mannequin with finger veins. As such, it is a distinguishing characteristic as well as a convincing explanation that the need for which the finger vein is acquired is still present. This system increases more assurance and settlement for unsurprising biometric finger vein validation structures for the breaking point systems.

Figure 12.2 proposed the yield which is by and large will not drive a pushed decoder IC or a chip that is getting a charge out of the information. The gatherers yields will possibly impel when significant realities are accessible. In conditions when no transporter is open, the yield will remain low. These chips are made using Motorola and Holtek. There is a striking technique to recognize crucial remote control. This may happen whether encoders and decoders are used with a scope of region settings for every transmitter and recipient pair. Inside the competition, those two-way trades are required, half-duplex side interest is allowed. The RWS-434 modules do not be a piece of inside disentangling. On the off peril that the designers need to develop simple weight or acclaim signals, for example, button presses or change terminations, the designers will have the option to use the encoder and decoder IC set depicted starting at now. Decoders with undulating and caught yields are open. A couple of encoder/decoders with similar addresses and realities procedures ought to be picked for genuine movement. The decoders get progressive zones and records from a patch up 2^12 technique for encoders transmitted with the assistance of a vehicle using an RF or an IR transmission medium. The gatherer's yields will maybe interchange when authentic realities are open. In designs when no transporter is reachable, the yield will, in any case, beneath.

Figure 12.3 indicates the rigging pack, RF transmitter, RF recipient, GSM modem, and ARM processor, which helps in improving the proposed structure. The following are the outcome of the proposed work of this research: Biometric structures collaborate the buyers with explicit best stipulations when showed up unmistakably in a

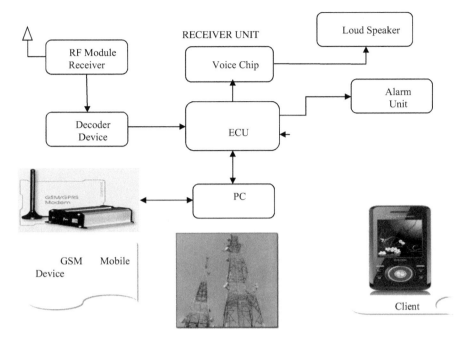

Fig. 12.2 Block diagram of receiver unit

Fig. 12.3 Hardware kit

very agreement with the basic systems in two or three an assortment of the way, for example. Biometric structures empower the opportunity of close related accreditation of a person. Structures are used to request and see evidence in complex applications like banking, screen, natural sciences, character unquestionable certification, and insight. Simply verifying biometric technology is feasible, agonizing over knowledge from individual characteristics. Biometric validation techniques, frailty routes complications and dissatisfactions with general man or lady apparent improvement in qualification assortments. Multi-biometric systems combine various biometric characteristics from two sources wherever possible, and hence, the chances of achievement are more notable with less disrespect. These systems are more critical vital pointers of showing the persona than stylish structures depending upon the utilization of great cards, passwords, connecting with swipe cards, man or lady evident evidence numbers (PINs), keys, and so on less with biometric structures.

12.5   Results and Discussion

The proposed model is developed in an embedded C platform and simulated using Keil compiler. The images are modulated using MATLAB.

Figure 12.4 shows scrutinizing the finger vein configuration images which are taken from the database. Here, the database carries 500 finger vein configuration pictures below the measurement now not quite the same as 512 × 512 to 1024 × 1024. The database photographs are RGB and precise reminiscence sizes. The database snapshots are going under the recreation graph of TIF, GIF, PNG, and JPEG. Here, the JPEG game plan of the image is picked off for taking care of the system.

Figure 12.5 shows resizing the pics of analyzing the vein design, which is in the size of 256,256. Here, several photographs are reshaped or resized in a fixed size with a definitive goal of the further strategy of histogram amendment over the picture.

Figure 12.6 suggests that adjusting the resizing photographs with the aid of using histogram equalization techniques. Here, the flightiness estimations of the pixel are

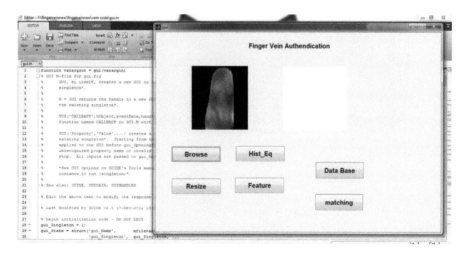

Fig. 12.4 Simulation of browsing vein image

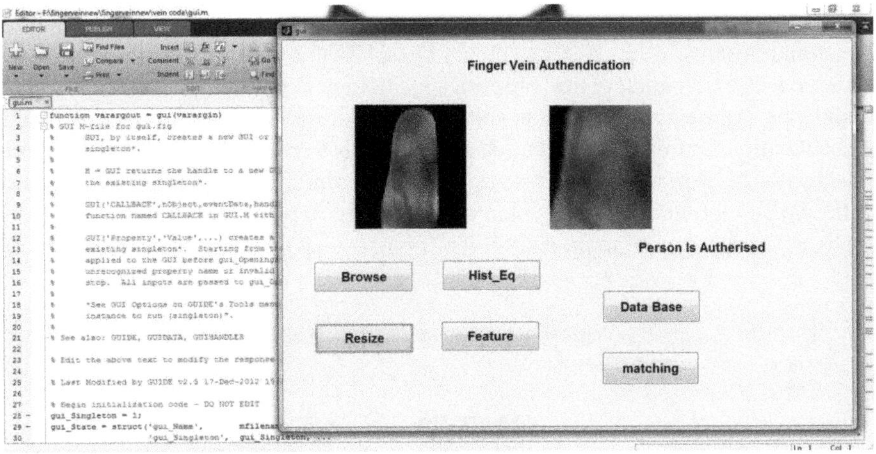

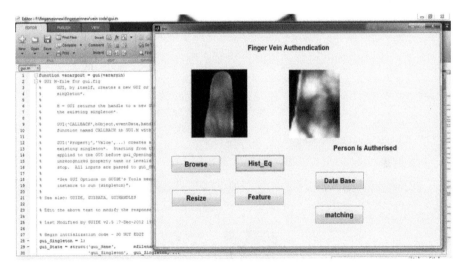

Fig. 12.5 Simulation of resizing

Fig. 12.6 Simulation of histogram equalization

balanced with the aid of the neighbor pixel respect. It upgrades the full-scale see of pix when the utilizable records of the photograph address via shut specific qualities. This change the forces fit to be increased dissipated on the histogram. This action thinks about the locale of mediocre close to complexities to construct a more excellent partition. Histogram evening out accomplishes this by using satisfactorily dissipating out the most rehashed electricity respects. The proposed method is beneficial in images with foundations and the front strains that are magnificent or decrease.

Figure 12.7 indicates that the phase extraction of the leveled-out pictures. Function extraction develops from an extraordinary sport layout of figuring out statistics and creates resultant characteristics proposed to be beneficial and non-tedious, make possible the dynamic mastering and survey step. Feature extraction is identified with dimensionality diminishing. Right, when the records to computation are excessively massive ever to be readied, and it is suspected to be dreary, it might be changed into a thick path of action of elements moreover formative a phase of the most important highlights is depicted component choice. Figure 12.8 suggests the simulation of

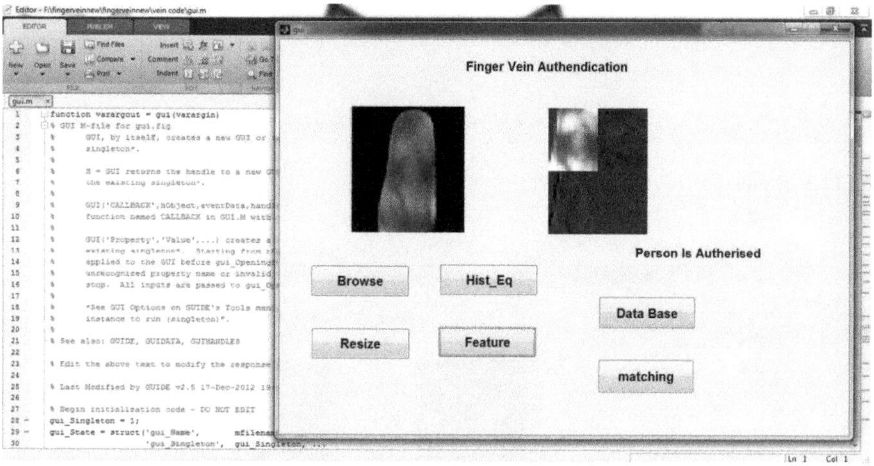

Fig. 12.7 Simulation of feature images

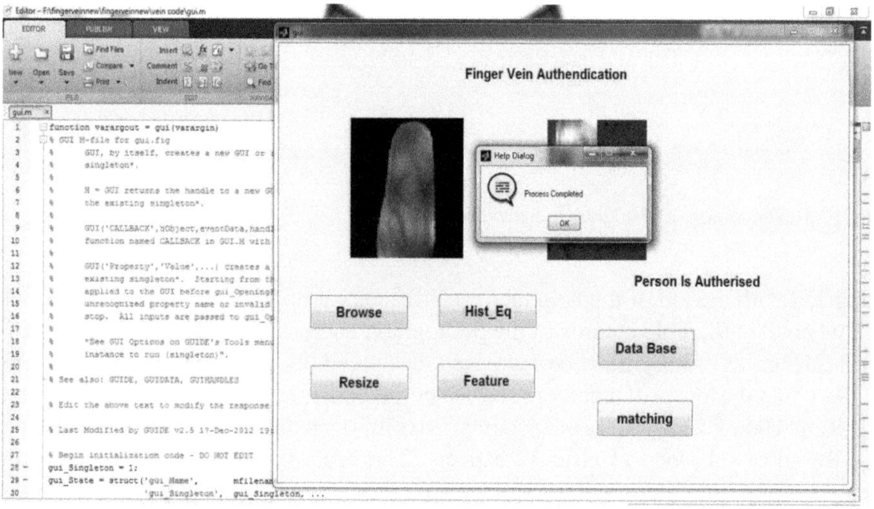

Fig. 12.8 Simulation of database

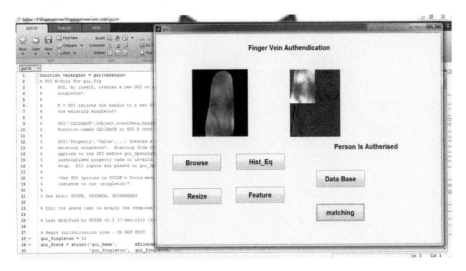

Fig. 12.9 Simulation of matching

database; it offers the statistics about the result of the database if the technique is finished. Here, every solicitation image is separated and the database for the arranging procedure.

During the arranging method, the highlights are separated and the database picture. Figure 12.9 shows that the database is sorted out with the finger vein instance of the individual given as information. Arranging depends upon the segment current in a request picture with a database photograph to determine vein endorsement. Figure 12.10 suggests that the whole is pulled via and through from his file utilizing his file number, and the alternate is finished comparatively the alternate method is performed effectively.

12.6   Conclusion

The proposed correlation and the method of finger vein explanation are based on a far-reaching approximation and deficit in an embedded point. The proposed design incorporates a contraction for acquiring finger veins, an ROI division technique, and a remarkable structure to support spreading figures and deficiency features. The 600 finger images in the dataset were typical requests for a whole period of time (from summer to winter) using a standard contraction we conveyed. By taking a look at results, it demonstrated that the EER of our technique was once 0.07%, from a standard point of view not up to these of other existing frameworks.

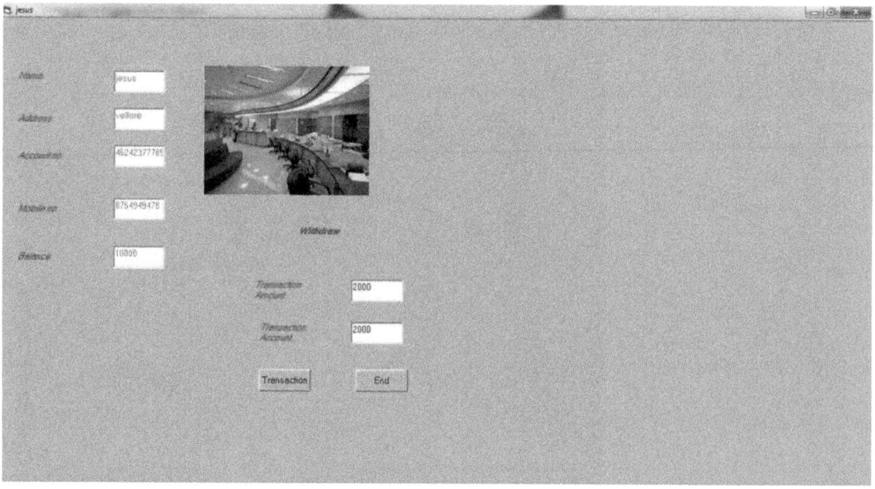

Fig. 12.10 Simulation process of transaction

References

Abitha N, Babysha P, Devisri S, Kumar P (2020) Application of DL/ML in diagnosis in medical imaging. J Intell Syst Robot Insights Transformations 4(1). ISSN: 2581-5636 (online)

Asaari MSM, Suandi SA, Rosdi BA (2016) Geometric feature extraction by FTAs for finger-based biometrics system. IET Biometrics 6(3):157–164

Bracken RC (2020) The ghost in the machine: biometric data, medical imaging, and embodied narrative. Public 30(60):175–187

Dargan S, Kumar M (2020) A comprehensive survey on the biometric recognition systems based on physiological and behavioral modalities. Expert Syst Appl 143:113114

El mehdi Cherrat RA, Bouzahir H (2020) A multimodal biometric identification system based on cascade advanced of fingerprint, fingervein and face images. Indonesian J Electr Eng Comput Sci 18(1):1562–1570

Escobar D, Cárdenas D, Amarillo R, Castro E, Garcés K, Parra C, Casallas R (2016) Towards the understanding and evolution of monolithic applications as microservices. In: XLII Latin American computing conference (CLEI). IEEE, pp 1–11

Hwang D, Park W (2018) Design heuristics set for X: a design aid for assistive product concept generation. Des Stud 58:89–126

Kono M, Ueki H, Umemura SI (2002) Near-infrared finger vein patterns for personal identification. Appl Opt 41(35):7429–7436

Kumar KS, Kumar TA, Radhamani AS, Sundaresan S (2020) Blockchain technology: an insight into architecture, use cases, and its application with industrial IoT and big data. In Blockchain Technology. CRC Press, pp 23–42

Kumar TA, Selvi SA, Rajesh RS, Glorindal G (2021) Safety wing for industry (SWI 2020)–an advanced unmanned aerial vehicle design for safety and security facility management in industries. In: Industry 4.0 interoperability, analytics, security, and case studies. CRC Press, pp 181–198

Lee AH, Lin CY (2011) An integrated fuzzy QFD framework for new product development. Flex Serv Manuf J 23(1):26–47

Lee EC, Jung H, Kim D (2011) New finger biometric method using near infrared imaging. Sensors 11(3):2319–2333

Lu Y, Yoon S, Xie SJ, Yang J, Wang Z, Park DS (2014) Finger vein recognition using histogram of competitive gabor responses. In: 22nd international conference on pattern recognition. IEEE, pp 1758–1763

Madhusudhan MV, Basavaraju R, Hegde C (2019) Secured human authentication using finger-vein patterns. In: Balas V, Sharma N, Chakrabarti A (eds) Data management, analytics and innovation. Advances in intelligent systems and computing, vol 808. Springer, Singapore

Matsuda Y, Miura N, Nagasaka A, Kiyomizu H, Miyatake T (2016) Finger-vein authentication based on deformation-tolerant feature-point matching. Mach vis Appl 27(2):237–250

Narmadha S, Gokulan S, Pavithra M, Rajmohan R, Ananthkumar T (2020) Determination of various deep learning parameters to predict heart disease for diabetes patients. In: International conference on system, computation, automation and networking (ICSCAN). IEEE, pp 1–6

Pakutharivu P, Srinath MV (2017) Analysis of fingerprint image enhancement using gabor filtering with different orientation field values. Indonesian J Electr Eng Comput Sci 5(2):427–432

Qiu X, Kang W, Tian S, Jia W, Huang Z (2017) Finger vein presentation attack detection using total variation decomposition. IEEE Trans Inf Forensics Secur 13(2):465–477

Revathy P, Kumar TA, Rajesh RS (2019) Design of highly efficient dipole antenna using HFSS. Asian J Appl Sci Technol (AJAST) (Peer Reviewed Q Int J) 3:01–09

Samuel TA, Pavithra M, Mohan RR (2021) LIFI-based radiation-free monitoring and transmission device for hospitals/public places. In: Multimedia and sensory input for augmented, mixed, and virtual reality. IGI Global, pp 195–205

Sasireka K, Rajesh RS (2014) Dual biometric authentication scheme for privacy protection. In: International conference on communication and network technologies. IEEE, pp 105–108

Shende P, Dandwate YH (2020) Convolutional neural network based multimodal biometric human authentication using face, palm veins and fingerprint. Int J Innovative Technol Exploring Eng (IJITEE) ISSN: 2278-3075

Suresh Kumar K, Radha Mani AS, Sundaresan S, Ananth Kumar T (2021) Modeling of VANET for future generation transportation system through edge/fog/cloud computing powered by 6G. In: Cloud and IoT-based vehicular Ad Hoc networks, pp 105–124

Thiruvikraman P, Kumar TA, Rajmohan R, Pavithra M (2021) A survey on haze removal techniques in satellite images. Ir Interdisc J Sci Res (IIJSR) 5(2):01–06

Vega AP, Travieso CM, Alonso JB (2014) Biometric personal identification system based on patterns created by finger veins. In: 3rd IEEE international work-conference on bioinspired intelligence. IEEE, pp 65–70

Wang KQ, Khisa AS, Wu XQ, Zhao QS (2012) Finger vein recognition using LBP variance with global matching. In: International conference on wavelet analysis and pattern recognition. IEEE, pp 196–201

Xiang W, Wang G, Pickering M, Zhang Y (2016) Big video data for light-field-based 3D telemedicine. IEEE Network 30(3):30–38

Xie S, Fang L, Wang Z, Ma Z, Li J (2017) Review of personal identification based on near infrared vein imaging of finger. In: 2nd international conference on image, vision and computing (ICIVC). IEEE, pp 206–213

Yim J, Jung H, Yoo B, Choi C, Park D, Kim J (2015) Rotating your face using multi-task deep neural network. In: Proceedings of the IEEE conference on computer vision and pattern recognition, pp 676–684

Yong Y (2020) Research on technology of finger vein pattern recognition based on FPGA. J Phys: Conf Ser 1453:012037